ELECTRIC POWER SYSTEMS RESILIENCY

ELECTRIC POWER SYSTEMS RESILIENCY

Modeling, Opportunity, and Challenges

Edited by

RAMESH C. BANSAL

Department of Electrical Engineering, University of Sharjah, Sharjah, United Arab Emirates; Department of Electrical, Electronic and Computer Engineering, University of Pretoria, Pretoria, South Africa

MANOHAR MISHRA

Department of Electrical and Electronics Engineering, FET, Siksha "O" Anusandhan University, Bhubaneswar, India

YOG RAJ SOOD

Jaypee Institute of Information Technology, Noida, India

Academic Press is an imprint of Elsevier
125 London Wall, London EC2Y 5AS, United Kingdom
525 B Street, Suite 1650, San Diego, CA 92101, United States
50 Hampshire Street, 5th Floor, Cambridge, MA 02139, United States
The Boulevard, Langford Lane, Kidlington, Oxford OX5 1GB, United Kingdom

ISBN 978-0-323-85536-5

For information on all Academic Press publications
visit our website at https://www.elsevier.com/books-and-journals

Publisher: Charlotte Cockel
Acquisitions Editor: Graham Nisbet
Editorial Project Manager: Sara Valentino
Production Project Manager: Anitha Sivaraj
Cover Designer: Mark Rogers

Typeset by STRAIVE, India

Contents

Contributors

Ramesh C. Bansal
Department of Electrical Engineering, University of Sharjah, Sharjah, United Arab Emirates; Department of Electrical, Electronic and Computer Engineering, University of Pretoria, Pretoria, South Africa

Javier Barrionuevo
Center for Studies of Energy Regulatory Activity, University of Buenos Aires (UBA), Argentina

Monalisa Biswal
Department of Electrical Engineering, National Institute of Technology, Raipur, India

Yendry González Cardoso
Universidad Autónoma de Nuevo León, San Nicolás de los Garza, Nuevo León, Mexico

Fanidhar Dewangan
Department of Electrical Engineering, National Institute of Technology, Raipur, India

Arturo Conde Enríquez
Universidad Autónoma de Nuevo León, San Nicolás de los Garza, Nuevo León, Mexico

Wilfredo C. Flores
Engineering Faculty, Universidad Tecnológica Centroamericana (UNITEC), Tegucigalpa, Honduras

Sonali Goel
Department of Electrical Engineering, Institute of Technical Education and Research (ITER), Siksha 'O' Anusandhan (Deemed to be University), Bhubaneswar, Odisha, India

Shazia Hasan
Department of Electrical and Electronics Engineering, Birla Institute of Technology & Science Pilani, Dubai Campus, Dubai, United Arab Emirates

Gowtham Kandaperumal
ComEd, Chicago, IL, United States

Subir Majumder
Lane Department of Computer Science and Electrical Engineering, West Virginia University, Morgantown, WV, United States

Manohar Mishra
Department of Electrical and Electronics Engineering, FET, Siksha "O" Anusandhan University, Bhubaneswar, India

Bignaraj Naik
Department of Computer Application, Veer Surendra Sai University of Technology, Burla, Odisha, India

Janmenjoy Nayak
Department of Computer Science, Maharaja Sriram Chandra Bhanja Deo (MSCB) University, Baripada, Odisha, India

Sanjoy Kumar Parida
Indian Institute of Technology, Patna, Bihta, India

Bhaskar Patnaik
Biju Patnaik University of Technology, Rourkela, Odisha, India

Danilo Pelusi
Faculty of Communication Sciences, University of Teramo, Teramo, Italy

Subhranshu Sekhar Puhan
Department of Electrical Engineering, Institute of Technical Education and Research (ITER), Siksha 'O' Anusandhan (Deemed to be University), Bhubaneswar, Odisha, India

Renu Sharma
Department of Electrical Engineering, Institute of Technical Education and Research (ITER), Siksha 'O' Anusandhan (Deemed to be University), Bhubaneswar, Odisha, India

Adhishree Srivastava
Birla Institute of Technology, Mesra, Ranchi, India

Anurag K. Srivastava
Lane Department of Computer Science and Electrical Engineering, West Virginia University, Morgantown, WV, United States

Santiago P. Torres
Department of Electrical, Electronics and Telecommunications Engineering, University of Cuenca (UC), Cuenca, Ecuador

Editors' biographies

Prof. Ramesh C. Bansal

Prof. Ramesh C. Bansal has more than 25 years of diversified experience in research, teaching and learning, accreditation, and industrial and academic leadership in several countries. Currently, he is a professor in the Department of Electrical Engineering at the University of Sharjah and extraordinary professor at the University of Pretoria, South Africa. Previously, he was employed at the University of Pretoria; University of Queensland, Australia; University of the South Pacific, Fiji; and BITS Pilani, India. He has published over 375 journal articles, conferences papers, books, and books chapters. He has more than 15,000 Google citations and an h-index of 57. He has supervised 25 PhD and 5 postdocs. His diversified research interests lie in the areas of renewable energy, power systems, and smart grids. Professor Bansal is an editor/associate editor for several journals including *IEEE Systems Journal* and *IET Renewable Power Generation*. He is a fellow and chartered engineer of IET (United Kingdom), fellow of the Institution of Engineers (India), and senior member of IEEE (United States).

Dr. Manohar Mishra

Dr. Manohar Mishra is an associate professor in the Department of Electronics and Electrical Engineering, under the Faculty of Engineering and Technology, Siksha "O" Anusandhan University, Bhubaneswar, Odisha, India. He has published more than 50 research papers in various reputed peer-reviewed international journals, conferences, and book chapters. He serves as an active reviewer for various reputed journals. He has more than 10 years of teaching experience in the field of electrical engineering. He is a senior member of IEEE and IEEE-PES Society. His area of interests include power system analysis and protection, signal processing, micro/smart-grid. He has served as convener and volume editor for several international conferences such as IEPCCT-2019, IEPCCT-2021, GTSCS-2020, and APSIT-2021. Currently, he serves as a guest editor for different journals such as *Neural Computing and Applications*, *International Journal of Power Electronics*, and *International Journal of Innovative Computing and Applications*.

Prof. Yog Raj Sood

Prof. Yog Raj Sood is currently working as Vice-Chancellor, Jaypee Institute of Information Technology, Noida. He has authored and coauthored several national and international publications and also works as a reviewer for reputed professional journals. He has an active association with different societies and academies around the world. He has received several awards for contributions to the scientific community. He has successfully completed three MHRD research projects and one mega project under TIFAC-CORE worth Rs. 53 million. Ten research scholars have already completed their PhD under his guidance. He is a senior member of IEEE, life member of the Indian Society for Technical Education (ISTE), fellow of the Institution of Engineers (India), and member of many other technical societies. His major research interests include deregulated power sector, congestion management, and microgrids.

Authors' biographies

Javier Barrionuevo
Universidad de Buenos Aires, UBA, Buenos Aires, Argentina
Email: j.barrionuevo@ieee.org

Professor Javier Barrionuevo received his master's degree in energy from Universidad de Buenos Aires, UBA. He graduated as Electrical Engineer from the Universidad Nacional de San Juan, Argentina. He is Professor in Centro de Estudios de la Actividad Regulatoria at the UBA and Consultant at Grupo Mercados Energeticos. Specializing in power systems, his area of interest includes the planning of the operation and expansion of the transmission systems and the integration of renewable energies.

Monalisa Biswal
Department of Electrical Engineering, National Institute of Technology, Raipur, India
Email: monalisabiswal22@gmail.com

Dr. Monalisa Biswal has received her MTech and PhD degrees from Veer Surendra Sai University of Technology, Burla, and Sambalpur University, Odisha, India, in 2008 and 2013. She has more than 6 years of industrial experience. Since 2013, she has been working as Assistant Professor in the Department of Electrical Engineering in the National Institute of Technology Raipur, Chhattisgarh, India. She is a recipient of POSOCO Power System Award (2014), DST-SERB Young Scientist Award (2015), and Chhattisgarh Young Scientist Award (2017). She is Senior Member of the Institute of Electrical and Electronics Engineers (IEEE), United States. She has published more than 90 journal articles and several chapters in books and has also presented papers at conferences. She has published a book titled *Microgrid: Operation, Control, Monitoring and Protection* (Springer Singapore, 2020). She is the principal investigator of three projects funded by SERB, (New Delhi, India) and EriGrid.

Arturo Conde Enríquez
Universidad Autonoma de Nuevo Leon, San Nicolás de los Garza, Mexico
Email: con_de@yahoo.com

Arturo Conde Enríquez received his MSc and PhD degrees in electric engineering from the Universidad Autónoma de Nuevo León, San Nicolás de los Garza, México, in 1996 and 2002, respectively. Currently, he is Professor with the Graduate Program of Electrical Engineering, Universidad Autónoma de Nuevo León. His research interests include adaptive protection of power systems, optimal energy management, and smart grid systems. Dr. Conde is a member of the National Research System of México.

Fanidhar Dewangan
Department of Electrical Engineering, National Institute of Technology, Raipur, India
Email: fanidhar.dewangan@gmail.com

Fanidhar Dewangan is currently working toward the PhD degree with the National Institute of Technology Raipur, Chhattisgarh. He received BE degree in electrical and electronics engineering in 2010 from Government Engineering College, Raipur. He has completed his MTech degree from DIMAT, Raipur, in 2018.

Wilfredo C. Flores
Universidad Tecnológica Centroamericana, UNITEC, Tegucigalpa, Honduras
Email: wilfredo.flores@unitec.edu.hn

Professor Wilfredo C. Flores received his PhD degree in electrical engineering from the National University de San Juan, Argentina; MBA from TEC of Monterrey, Mexico; and BEE from the National University of Honduras, UNAH. Dr. Flores works as Researcher and Professor in Universidad Tecnologica Centroamericana, UNITEC, from Honduras. His areas of interest cover topics such as analysis of power systems, regulatory framework of power markets, energy policy, power transformers, and computational intelligence.

Sonali Goel
Department of Electrical Engineering, ITER, Siksha 'O' Anusandhan Deemed to be University, Bhubaneswar, India
Email: subhranshusekharpuhan@soa.ac.in

Dr. Sonali Goel has completed her PhD degree in the field of solar photovoltaic systems from the Institute of Technical Education and Research, Siksha 'O' Anusandhan Deemed to be University, Bhubaneswar, Odisha in the year 2018. Her PhD research work is on the topic "Design and Performance Characteristics of Solar Photovoltaic and Biomass Hybrid Energy Systems for Rural Application." Her research interest includes solar photovoltaic systems, biomass and biogas systems, hybrid renewable energy system, and rural electrification and electric vehicles. She has been the Life Member of Solar Energy Society of India, Associate Member Institution of Engineers (India), Life Member of Indian Society of Lighting Engineers (India), and Student Membership of IEEE. She has been the author of more than 10 articles in international journals published by Elsevier, Springer, and Taylor and Francis; 8 book chapters; and 20 conference papers of national and international repute.

Yendry González Cardoso
PhD student of the Universidad Autonoma de Nuevo Leon, San Nicolás de los Garza, Mexico

Email: yendryg1985@gmail.com

Yendry González Cardoso received his BEng and MSc degrees in electrical engineering from the Central University Marta Abreus of the Villas (UCLV), Villa Clara, Cuba, in 2009 and 2017, respectively. He is pursuing his PhD in electrical engineering from the Universidad Autonoma de Nuevo Leon (UANL), San Nicolás de los Garza, México. His research interests are power system protection, distributed generation, and smart grid systems.

Shazia Hasan
Department of Electrical and Electronics Engineering, Birla Institute of Technology & Science, Dubai Campus, Dubai, United Arab Emirates

Email: shazia.hasan@dubai.bits-pilani.ac.in

Shazia Hasan was born in India in 1980. She received her bachelor's degree in electronics and telecommunication from UCE, Burla, India, in 2002, and PhD in engineering from Biju Patnaik University of Technology, India, in 2012. She has more than 15 years of teaching experience in different engineering colleges. Currently, she is Assistant Professor at BITS Pilani Dubai Campus from 2015. Dr. Shazia Hasan won "Young Scientist Award" from VIFRA in 2015. She is the author of several papers in peer-reviewed international/national journals and international/national conferences. She also served as Reviewer of international journals. She organized several conference/seminars/workshops in the past. She successfully completed the industrial consultancy project from "Star Cement" as Principal Investigator. Her research interest includes digital signal processing, biomedical signal processing, and statistical filter design.

Gowtham Kandaperumal
School of Electrical Engineering and Computer Science, Washington State University, Pullman, WA, United States
Email: g.kandaperumal@wsu.edu

Gowtham Kandaperumal (Member, IEEE) received his master's degree in electrical engineering from Arizona State University in 2014, and PhD degree in electrical engineering and computer science from Washington State University in 2021. He is Senior Engineer with the Reliability Group, Commonwealth Edison (ComEd). Prior to ComEd, he was Distinguished Graduate Research Fellow with Pacific Northwest National Laboratory from 2019 to 2021. He was an electrical engineer with Affiliated Engineers, Inc., from 2014 to 2017.

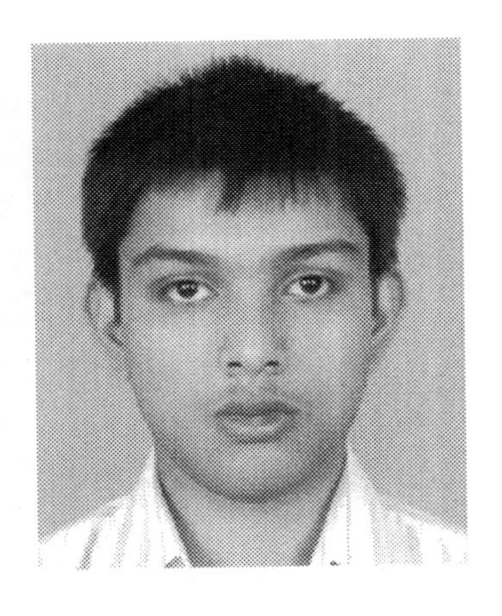

Subir Majumder
Lane Department of Computer Science and Electrical Engineering, West Virginia University, Morgantown, WV, United States
Email: subir.majumder@mail.wvu.edu

Subir Majumder received the PhD degree under a Cotutelle/Joint Agreement between Indian Institute of Technology Bombay, India and University of Wollongong, Australia in 2020. From 2020 to 2021, he worked as a postdoctoral research associate at the Washington State University, Pullman, WA, United States. Currently, he is working as an Engineering Scientist at the Lane Department of Computer Science and Electrical Engineering, West Virginia University, Morgantown, WV, United States. He was conferred POSOCO Power System Awards (PPSA) under Doctoral category in 2020. His research interests include power systems modeling, operations (including operational resiliency) and planning, power system economics, distributed optimization, power quality and the smart grid.

Bignaraj Naik
Department of Computer Application, Veer Surendra Sai University of Technology, Burla, India
Email: mailtobnaik@gmail.com

Bighnaraj Naik (M'19) is Assistant Professor in the Department of Computer Application, Veer Surendra Sai University of Technology (Formerly UCE Burla), Odisha, India. He received his PhD in computer science and engineering, MTech in computer science and engineering, and BE in information technology in 2016, 2009, and 2006, respectively. He has published more than 90 research papers in various reputed peer-reviewed international journals, conferences, and book chapters. He has edited four books from various publishers such as Elsevier, Springer, and IGI Global. At present, he has more than 10 years of teaching experience in the field of computer science and IT. His area of interest includes data mining, computational intelligence, and soft computing and its applications. He has been serving as Guest Editor of various journal special issues from Elsevier, Springer, and Inderscience.

Janmenjoy Nayak
Department of Computer Science, Maharaja Sriram Chandra Bhanja Deo (MSCBD) University, Mayurbhanj, Odisha, India
Email: mailforjnayak@gmail.com

Janmenjoy Nayak (M'19) is working as Assistant Professor, Department of Computer Science, Maharaja Sriram Chandra Bhanja Deo (MSCBD) University, Mayurbhanj, Odisha, India. He has published more than 100 research papers in various reputed peer-reviewed referred journals, international conferences, and book chapters. His area of interest includes data mining, nature-inspired algorithms, and soft computing. He is the regular member of IEEE and life member of some of the reputed societies such as CSI India. He has edited nine books from various publishers such as Elsevier and Springer. He has been serving as Guest Editor of various journal special issues from Elsevier, Springer, and Inderscience. He has served as Volume Editor of Series International Conference on Computational Intelligence in Data Mining, Computational Intelligence in Pattern Recognition, and so on.

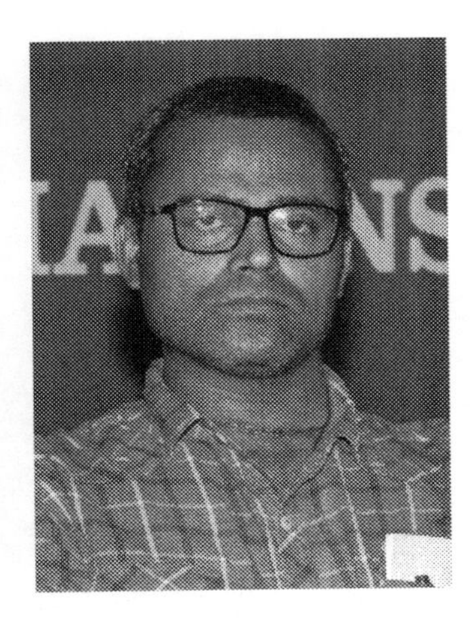

Sanjoy Kumar Parida
Associate Professor, Electrical Engineering Department, IIT Patna, Bihta, India
Email: skparida@iitp.ac.in

K. Parida (SMIEEE, 2016) is currently Associate Professor in the Department of Electrical Engineering, IIT Patna. He was awarded the Senior Research Fellowship (SRF) by the Power Management Institute, NTPC Ltd., Noida, in 2007, during his PhD program. He has received Bhaskar Advanced Solar Energy (BASE-2015) Research Fellowship award by Indo US Science and Technology Forum (IUSSTF), DST, Government of India in 2015, and Young Faculty Research Fellowship (YFRF) award by Digital India Ltd., Ministry of Electronics and Information Technology (MeitY), Government of India in 2019. His research interest includes power system protection, control, and stability.

Bhaskar Patnaik
Department of Electrical Engineering, under the Gandhi Institute of Excellent Technocrats, Biju Patnaik University of Technology, Rourkela, India
Email: bhaskar7310@gmail.com

Dr. Bhaskar Patnaik is Associate Professor in the Department of Electrical Engineering under the Gandhi Institute of Excellent Technocrats, BPUT, Odisha. He has more than 20 years of teaching experience in the field of electrical engineering. His area of interest includes power system analysis and protection, signal processing, and micro/smart-grid.

Danilo Pelusi
Faculty of Communication Sciences, University of Teramo, Teramo, Italy
Email: dpelusi@unite.it

Danilo Pelusi received his PhD degree in computational astrophysics from the University of Teramo, Italy. He is working as an Associate Professor in the Department of Communication Sciences, University of Teramo; Editor of books published by Springer and Elsevier; Associate Editor of *IEEE Transactions on Emerging Topics in Computational Intelligence* (2017–20), *IEEE Access* (2018–present), and *International Journal of Machine Learning and Cybernetics*, Springer (2019–present); and Guest Editor for Elsevier, Springer, MDPI, and Hindawi journals. He is the editorial board member of many journals. He is the keynote speaker and chair of IEEE conferences and inventor of patents on artificial intelligence. His research interests include fuzzy logic, neural networks, information theory, machine learning, and evolutionary algorithms.

Subhranshu Sekhar Puhan
Department of Electrical Engineering, ITER, Siksha 'O' Anusandhan Deemed to be University, Bhubaneswar, India
Email: subhranshusekharpuhan@soa.ac.in

Mr. Subhranshu Sekhar Puhan has completed his MTech degree in the specialization of power system from IIT Delhi, in 2016. Currently, he is working as Assistant Professor in the Department of Electrical Engineering, Institute of Technical Education & Research, Siksha 'O' Anusandhan Deemed to be University, Bhubaneswar. His research interest includes novel controller design for solar photovoltaic system, hybrid renewable energy system, nonlinear controller application in modern power system, and design and control of electric vehicles. He has been Student Member of IEEE and is currently pursuing his PhD. He has been the author of four conference papers in international conferences in IEEE and two book chapters in Springer.

Renu Sharma
Department of Electrical Engineering, Institute of Technical Education and Research, Siksha 'O' Anusandhan Deemed to be University, Bhubaneswar, India
Email: renusharma@soa.ac.in

Renu Sharma is working as Professor and Head of the Department of Electrical Engineering in Siksha 'O'Anusandhan Deemed to be University, Bhubaneswar, India. She is Senior Member of IEEE, Life Member of IE (India), Member of IET, Life Member of ISTE, Life Member of ISSE, Past Chair of WIE IEEE Bhubaneswar subsection. Her research areas are smart grid, soft computing, solar photovoltaic systems, power system scheduling, evolutionary algorithms, and wireless sensor networks. She has published around 100 journal and conference articles of international repute. She has organized several national and international conferences. She is Guest Editor of a special issue of the *International Journal of Power Electronics*, Inderscience. She is also Guest Editor of a special issue of the *International Journal of Innovative Computing and Applications*, Inderscience. She has around 21 years of leading impactful technical, professional, and educational experience. She was holding the position of General in IEEE ODICON 2021, flagship conference IEEE WIECON-ECE 2020, Springer conference GTSCS-2020, IEPCCT-2019, IEPCCT 2021.

Adhishree Srivastava
Assistant Professor, Electrical and Electronics Engineering Department, BIT Mesra, Ranchi, Patna Campus, India
Email: adhishree@bitmesra.ac.in, adhishree56@gmail.com

Adhishree Srivastava received her BE degree in electrical and electronics engineering from the Birla Institute of Technology Mesra, Ranchi, India, in 2012, and MTech degree with a gold medal in power system engineering from the

Motilal Nehru National Institute of Technology, Allahabad, India, in 2014. She worked as Edison Engineer in General Electric Private Ltd. from 2014 to 2016. Currently, she is working as Assistant Professor in BIT Mesra, Patna, Bihar, India. She is also pursuing her PhD in electrical engineering at the Indian Institute of Technology, Patna, India. She has authored various papers in international journals and conferences. Her research interests include optimal relay coordination, distributed generation, and microgrid protection using machine learning and deep learning.

Anurag K. Srivastava

Lane Department of Computer Science and Electrical Engineering, West Virginia University, Morgantown, WV, United States

School of Electrical Engineering and Computer Science, Washington State University, Pullman, WA, United States

Email: anurag.srivastava@mail.wvu.edu

Anurag K. Srivastava (Senior Member, IEEE) received his PhD degree in electrical engineering from the Illinois Institute of Technology, in 2005. He is currently Lane Professor and Chairperson of the Lane Department of Computer Science and Electrical Engineering, West Virginia University, and Adjunct Professor of electric power engineering at Washington State University. He also has a joint appointment as Senior Scientist with the Pacific Northwest National Laboratory (PNNL). He has authored more than 300 technical publications, including a book on power system security and four patents. His research interest includes data-driven algorithms for power system operation and control, including resiliency analysis. He is serving or served as Editor for the *IEEE Transactions on Smart Grid*, *IEEE Transactions on Power Systems*, and *IEEE Transactions on Industry Applications*.

Santiago P. Torres
Universidad de Cuenca, Cuenca, Ecuador
Email: santiago.torres@ucuenca.edu.ec

Dr. Torres is currently the Head of the Electrical, Electronics, and Telecommunications Department at the University of Cuenca, in Ecuador. His research topics are related to electric transmission systems, electric expansion planning, stability and security, and AI applications to power systems.

Foreword

Prof. (Dr.) Girish Kumar Singh
Department of Electrical Engineering, Indian Institute of Technology Roorkee, Roorkee, Uttarakhand, India
Email: girish.singh@ee.iitr.ac.in

Resiliency of power systems is gaining increasing attention from researchers and corporate entities largely because of a growing number of outages due to climate change-induced disastrous events, which are also showing an increasing trend of occurrence in recent times. At the same time, the radical change in the demands and expectations of performance of power systems in terms of 21st-century parameters, coupled with consumers' participation through distributed generation and need for green energy, has brought about a paradigm shift in the way the new age power system network have evolved. Ensuring uninterrupted power supply to all the customers connected to the utility by mitigating and/or overriding frequently occurring temporary and permanent system faults is what preserving the reliability of the system is all about. A resilient system on the other hand should prevent long duration outages caused by low-frequency high-impact (LFHI) events; it must be able to predict, prepare for, endure, and bounce back to normalcy before, during, and after the occurrence of a severely disruptive event. As these so-called LFHI events are rare and unpredictable, a resilient system must also learn from its past experiences and adapt itself to be resilient against the reoccurrence of such events.

Historically the focus of electrical engineers in power system research has been highly reliant on traditional methods based on system parameter measurement and analysis. But in the last decade, there has been a notable rise in research concerned with power system fault detection and assessment using advanced relaying systems based on artificial intelligence (AI), Internet-of-Things (IoT), machine learning (ML), cloud computing, and cyber physical systems. Most of this research has revolved around predicting system stability and short-circuit faults, reliability analysis, and grid security, but researchers are also interested in risk factors predictive of system conditions. There have been several different approaches to enhance power system resiliency prescribed in literatures; a more generalized resilience evaluation, quantification, and economic realization is still awaited and leaves ample scope of study for future researchers.

This book's chapters present the very best in fundamentals and application perspectives to deal with the issue of resiliency in power system. Topics run the gamut of research in this field: overview and architecture of resilient energy systems highlighting its issues and challenges, remedial action scheme to increase resiliency under failures in the power grid, enhancing relay resiliency in an electric power transmission system, power system stability data analysis using ML approaches, microgrids as a resilience resource with a focus on real-world applications, IoT and fog computing application to improve smartgrid resiliency, application of artificial neural network for load forecasting in an electric power system to increase resiliency, and cost benefit analysis for smart grid resiliency.

Because of the accelerating research progress in IoT, AI, ML, and cyber security and its use in electric power systems in all its complexity, this book's publication is not only timely but also much needed. This book contains 10 peer-reviewed chapters reporting state of the art in resiliency research in power systems as it relates to the most fundamental aspects as well as the application of AI and computational intelligence techniques, covering many important topics in contemporary prognostics. In my opinion, this book will be a valuable resource for graduate students, engineers, and researchers interested in understanding and investigating this important field of study.

Preface

With climatic change, the world is witnessing increased instances of extremely devastating climatic events apart from many other low-frequency high-impact calamities, which leave behind trails of massive destruction. These calamities take a heavy toll on the electrical power systems (EPSs) infrastructures, leading to power outages for sustained periods before the infrastructure is remedied and the system creeps back to recovery. The modern world and its way of life as well as means of livelihood are heavily dependent upon continuous and reliable power supply. Outages are unwarranted and considered a serious threat to the economy and development of a nation. In such circumstances, the power system needs to be resilient against these disruptive events. Resiliency of power systems is now viewed as a matter of great concern as investment toward making a system resilient is thought to be more prudent than trying to bring down the price of power. The resiliency of an EPS is defined as its ability to bounce back to equilibrium (stable operating point) after a major disruptive event. "Robustness," "resourcefulness," "rapid recovery," and "adaptability" are the four features considered to be the attributes of resilience by the National Infrastructure Advisory Council (NIAC), United States. Rapid recovery and adaptability to an unprecedented disruption are the major requirements for ensuring resiliency of an EPS, in which a microgrid is considered the most viable solution in this aspect. The presence of renewable energy sources, their ability to feed critical loads during system contingencies through islanded mode of operation, operational flexibility, and self-healing capabilities are the features that qualify microgrids as most suitable and effective resilient resources for service restoration in the event of "low probability high impact (LPHI)" events. "Resiliency of power systems" is in the nascent stage of research. The actual meaning of the term "resiliency" has been defined in several ways in various research works; however, a proper and unifying definition of this term is yet to be confirmed by researchers.

This book aims to provide resilience and associated metrics that can be used quantitatively. Moreover, several methodologies related to improvement of the power system's resilience through optimal selection and location of infrastructure as well as resources, high penetration of renewable energy resources, cyber physical systems, Internet of Things, Internet of Services, and advanced computing techniques have been included. A brief description of the topics covered follows.

Chapter 1 provides a detailed analysis on the overview and architecture of resilient energy systems. The need for resiliency improvement in the EPS, the difference between resiliency and reliability, and the importance of resiliency improvement are discussed and descriptions of associated resiliency matrices are presented.

Chapter 2 provides a comprehensive review of resilient smart-grid systems. It describes the issues and challenges of smart grids with the aim of identifying means to improve the resiliency. Moreover, it analyzes smart-grid resilience matrices and different methods for resiliency improvement.

Chapter 3 presents the implementation and simulation results of a remedial action scheme applied to the Central American electric power grid. It is proven that the implementation of this regional protection will increase network resiliency under failures that could cause regional blackouts. Four case studies that show the improvement of the Central American power grid performance are analyzed.

Chapter 4 discusses the coordination of directional overcurrent relays (DOCRs) with the application of metaheuristic methods. The coordination of DOCRs is solved by formulating the optimization problem based on the fulfillment of time constraints. The proposed coordination method reduces the operation time of the relays compared with the traditional coordination method. The use of nonconventional curves is considered to provide the flexibility needed to meet the time intervals at each required location to improve relay resiliency in an EPS.

Chapter 5 presents the stability issues of smart-grid cyber-physical systems (CPSs) and the computational role of intelligence techniques in dealing with the issues. An improved genetic algorithm (GA)-based extreme learning machine (ELM) model for smart-grid CPS stability prediction is presented. The results of the suggested model are compared with other profound computational intelligence models.

Chapter 6 discusses the challenges of conventional protection schemes applied to the current renewable penetrated power system. Different protection schemes in active distribution networks and microgrids are presented. The major issues related to microgrid protection are presented and various possible solutions are discussed comprehensively. Some of the recent machine learning (ML)-based schemes to perform protection tasks such as fault detection, fault type, and faulty phase classification and location are presented.

The application of data-driven and ML-based methodology is proving to be very efficient in maintaining the resiliency of an active grid.

Chapter 7 discusses the strategies and concepts that enable resiliency through microgrids. Existing research deals with the formation and control of microgrids with a tendency to focus on a broader concept of system survival during an LPHI event. However, there is no solid approach developed for engineers and researchers to use the resourcefulness of microgrids to serve as a resiliency asset during an LPHI event. This chapter helps readers to understand the key features of microgrid resiliency, evaluation of resiliency and the challenges, and the future of microgrids as a resiliency resource.

Chapter 8 presents the use of IoT and Fog computing in smart-grid applications. Different applications of IoT technologies in smart grids to improve the security issues and resiliency are presented. The chapter investigates the current status of Fog computing applications in smart-grid ecosystems with the integration of IoT technology. It also elaborates on the related security issues and challenges of IoT and Fog computing in smart grid systems.

Chapter 9 discusses the role of advanced forecasting techniques in dealing with the resiliency issue. A neural network-based load forecasting method is presented for proper energy management. The effects of load forecasting uncertainties are also studied from a power system reliability point of view.

Chapter 10 states that the ability of the power system to withstand, respond to, and recover from a catastrophic event is an important factor often used to define the resilience of a power system. Environmental threats and human threats, such as cyber security attacks, may trigger these types of incidents. Cost-benefit analysis (CBA) is a proven method to assess the economic feasibility of development interventions. The cost of undertaking any project can be compared by using CBA and by knowing their net benefit and efficiency. A flexible framework for CBA can assist in assessing and prioritizing investments to improve the resiliency of the energy system. This chapter deals with the economic approach to calculate the benefit and cost of any project. CBA should be seen as a preferred choice by proving that benefit outweighs the cost and providing significance to the community.

We thank all the contributors and the reviewers for their valuable contributions and dedicated efforts toward the successful completion of this book. We thank the editorial team at Elsevier for their valuable technical support and timely publication of the book. We thank our universities, authorities, and staff members for maintaining a cordial atmosphere and providing the facilities for the completion of the book. We express gratitude and sincere regards to our family members who have provided great support during preparation of this book.

Ramesh C. Bansal, Editor
Manohar Mishra, Editor
Yog Raj Sood, Editor

CHAPTER ONE

Overview and architecture of resilient energy systems

Bhaskar Patnaik[a], Manohar Mishra[b], and Ramesh C. Bansal[c,d]

[a]Biju Patnaik University of Technology, Rourkela, Odisha, India
[b]Department of Electrical and Electronics Engineering, FET, Siksha "O" Anusandhan University, Bhubaneswar, India
[c]Department of Electrical Engineering, University of Sharjah, Sharjah, United Arab Emirates
[d]Department of Electrical, Electronic and Computer Engineering, University of Pretoria, Pretoria, South Africa

1. Introduction

The introduction to power system resiliency can be preluded with the following excerpt from a recent news article in Bloom Energy (Frank, 2019): "As outages increase, business are considering the 'cost of not having power' instead of just the 'cost of power.' Energy resilience is becoming an issue business leaders can no longer afford to neglect-both from a strategic and cost perspective." The article, while highlighting the energy trends in 2020 and in coming years, predicts that escalating climate challenges and vulnerability of the grid to the devastating effects of these extreme climatic conditions have elevated the focus of business leaders on energy resiliency and sustainability. The losses incurred because of increased number of outages in recent past have far reaching negative consequences than that of managing the cost of power generation. Electric companies in the United States have reported more than 2500 major outages since 2002 and nearly half of these (1172 to be specific) were caused by weather conditions, such as storms, hurricanes, and other unspecified severe weather conditions (Hussain & Pande, 2020). Ironically, such incidences are having an increasing trend. Depending upon the geographic location, each and every country suffers from the vagaries of nature, which leaves behind massive infrastructural devastation. And power infrastructure always remains the major casualty with significant loss in terms of economy and sufferings inflicted upon people.

Electric Power Systems Resiliency
https://doi.org/10.1016/B978-0-323-85536-5.00007-2

2. Understanding the causes of outages

There are a wide range of events which can lead to power outages, climatic conditions being just one of them. In terms of frequency of occurrence and quantum of devastation inflicted, these events are sometimes categorized as "low-frequency high-impact" (LFHI) and "high-frequency low-impact" (HFLI) events. Needless to say, tackling LFHI events are the most challenging tasks. LFHI events can also be termed as "large area long duration" (LALD) events considering the vastness of the affected area and longevity of the ensuing power outage. Generally, these events are unpredictable with regards to their time of happening and the quantum of devastation. However, certain technological developments in weather and disaster forecasting do provide some warning period to face certain types of eventualities (Enhancing the Resilience of the Nation Electricity System, 2017). Apart from extreme weather conditions, there are many factors which leads to outages, such as cyberattack leading to power system communication failure and subsequent disruption, political misgivings leading to discontinuation of fuel supply chain, extreme climatic condition (other than weather) ensuing intentional outages for particular city or zone, etc. The nature of disruptive events varies widely in terms of predictability, size of the affected geographical area, and availability of warning time to prepare for the eventuality. The extent of the impact of a disruptive event depends on the existing condition of the power system infrastructure, interdependence of the infrastructures, and the spatial and temporal impacts of the disruptive event itself. A comprehensive list of all such possible events with tentative warning periods and their impact on power system resiliency are provided in Table 1.1.

3. Why resiliency?

Irrespective of the causes, one thing that is of growing concern is that outages translate in loss to economy and uncouth misery to the people who have become ever dependent on electricity for even a bare minimum comfort of a lifestyle. In these contexts, it has become imperative that EPSs worldwide should be resilient enough. A resilient EPS must ensure uninterrupted power supply overriding the temporary insignificant fault conditions and withstanding the numerous disruptive and destructive calamities or events as enlisted before. In short, it must be robust enough to be

Table 1.1 Characteristics of low-frequency high-impact (LFHI) extremely disruptive events.

LFHI events/threats (to electrical utility grids and services)	Expected advance warning time (preoccurrence of the LFHI event)	Required recovery time (postoccurrence of the LFHI event)	Extent of damage/consequence/impact	Instances
Cyberattack	Limited or no warning (can be almost instantaneous). Sometimes, the hacker may be collecting information covertly while seemingly remaining inactive for months	Duration of restoration is unclear and entirely depends on the nature of attack	Failure in preventing the intrusion of malware to critical systems may jeopardize the entire system. Though physical damages do not happen, but there is always a possibility of so depending upon the sophistication of the attack. Modern power systems which rely heavily on software and communication networks are most vulnerable to this attack, such as SCADA, power plant distributed control systems, smart grid technologies, distributed energy resources, etc.	Cyberattack on Ukrainian power system (2015)
Drought and associated water shortage	Sufficient warning time gets available		Have manyfold impact on power systems. Reduced hydroelectricity generation, scarcity of cooling water for cooling towers of power stations, and increase in demand of power for pumping purpose are few of the implications of drought leading to stress on power system. Hardware failures and error by operators due to stress on power system can result in increased significant outages	

Continued

Table 1.1 Characteristics of low-frequency high-impact (LFHI) extremely disruptive events—cont'd

LFHI events/threats (to electrical utility grids and services)	Expected advance warning time (preoccurrence of the LFHI event)	Required recovery time (postoccurrence of the LFHI event)	Extent of damage/consequence/ impact	Instances
Earthquake	Happens without warning. Possible to obtain several seconds of advance warning by recording the propagation velocity of the earthquake using sophisticated instruments	Restoration following a massive earthquake is a massive task and take quite a long time in terms of days and weeks	Significant potential of disruption of major power system equipment, including distribution poles, transmission towers, substations, and generating units. Possibility of damage to natural gas systems due to vulnerability of supply pipelines to earthquakes	Few major earthquakes of recent past: 2011_Sendai_Japan 2010_Bio-Bio_Chile 2004_Sumatra_Indonesia
Flood/storm surge	Floods can be of many forms. While flash floods happen unexpected, the buildup of floods because of hurricanes and tropical storms do provide adequate warning period	Flood mapping using historical flood data helps choosing the locations least likely to be inundated for placing major facilities of the utility grid. However, climatic changes may put the historic data-based flood mapping go wrong. This requires careful planning and design of underground electrical systems in areas prone to frequent flood	Floods can cause serious damage to distribution and transmission towers, equipment installed on the ground, and weaken the footings of the poles and towers. Flooding makes access to distribution system difficult for the repair crews	History is full of instances of floods leaving trails of massive destruction of infrastructure as well as toll of people. Few recently happened floods can be listed out as: 2016_India_Flood caused by Monsoon Rain 2013_North India Floods 2011_Southeast Asian Floods

Hurricanes or tropical cyclones	The vastly improved present day weather forecasting models can provide warning several days in advance. Intensity and location of landfall of the storms or cyclones can now be predicted with high degree of accuracy	Recovery from hurricane may take several days or weeks depending upon the extent of damage to the power system infrastructure	Tropical cyclones can cause extensive damage to the power systems. Cyclones have threefold impacts on the power system infrastructure. High-speed winds on the land, followed by inundation of coastal areas due to storm surge and flooding due to precipitation	Super cyclone of 1999, Filine (2011), Hudhud (2012), Foni (2019), all in India in recent times
Ice storm	Season specific, hence adequate preventive measures can be and is taken well in advance	With well proven emerging techniques, risk from ice storms could be minimized and recovery time can be accelerated. Complete recovery may take weeks	Can cause widespread damage to distribution and transmission line systems because of collapse of poles and structure on accumulation of ice	1998_South-eastern Canada and North-eastern United States
Major operation error	Blackouts due to operational error can take place over a period of few minutes to hours, providing the opportunity to detect the error and take appropriate corrective actions	It may take minutes to days to recover the system depending on the cause and nature of fault which initiated the operational error	Blackouts or cascading blackouts are the consequences of major operational errors, though no serious damage to the physical pats of the power system is expected	Northeast, United States (2003) blackout. Southwestern, United States (2011) blackout

Continued

Table 1.1 Characteristics of low-frequency high-impact (LFHI) extremely disruptive events—cont'd

LFHI events/threats (to electrical utility grids and services)	Expected advance warning time (preoccurrence of the LFHI event)	Required recovery time (postoccurrence of the LFHI event)	Extent of damage/consequence/impact	Instances
Physical attack	Can occur without or with limited warning	Recovery may take many days or weeks	Possibilities of significant damage to important system components, such as large transformers, substations, and equipment cannot be ruled out	Bombing and terrorist activities and such as those happened in Afghanistan, Colombia, Peru, and Thailand, to cite a few (NRC 2012). Planned sniper attack on Metcalf transmission substation (2013) in the United States
Regional storms and tornadoes	Occurrence and frequency of events like tornadoes are unpredictable, while storms and few other regional weather-related events can be predicted based on scientific understanding of these not so rare events	With planned preparedness, the locally affected systems can be repaired and the system can be recovered within few hours to days	Damage is limited to specific regions and are not widespread. However, the extent of damage could be massive depending upon the type and intensity of weather event	Tornado affecting Mississippi to North Carolina, United States (2006)

Space weathers and other electromagnetic threats	30 min of advance warning by satellites and sun observation up to 2–3 days ahead of impact can be expected, providing opportunity to protect the grid through implementation of operating procedures that are designed to protect critical components of the EPS. The time constants determining impacts on transformers from solar storms (or from the E3 portion of electromagnetic pulse (EMP) events) are slow enough and hence transformers, which form the major component of the EPS, can be protected during the occurrence of the event	A real-time monitoring through a well-developed standard of approach to mitigate the impact of solar storms on transformers is highly desirable. Such real-time monitoring combined with automated protection schemes can further help prevent damage to transformers due to geomagnetic disturbances	There are varieties of solar activities, also known as space weather, which can affect earth's environment. Coronal mass ejections can cause fluctuations in earth's magnetic field leading to creation of very low-frequency voltage gradients across land. The voltage gradient induces quasisteady state current in long transmission lines, causing magnetic saturation of the transformer cores, which results in overheating of the core and subsequent damage of the transformers	1859_Carrington: Failure of United States and Europe telegraph systems,1989_Québec: Power supply disruption for almost 9 h affecting over 6 million people. A smaller hour-long outage occurred in Sweden in October 2003

Continued

Table 1.1 Characteristics of low-frequency high-impact (LFHI) extremely disruptive events—cont'd

LFHI events/threats (to electrical utility grids and services)	Expected advance warning time (preoccurrence of the LFHI event)	Required recovery time (postoccurrence of the LFHI event)	Extent of damage/consequence/impact	Instances
Tsunami	Impact of tsunami is limited to coastal areas, and with advanced warning systems, sufficient time can be obtained to shut down the critical facilities to minimize the damage	Restoration and recovery may take days or weeks depending upon the magnitude of damage to the electricity infrastructure	With proper utility planning, the effect of tsunami on electrical power systems can be significantly curtailed. Existing facilities located in tsunami vulnerable coastal areas however suffer major devastation	Puerto Rico, United States (1918) Alaska, United States (1946). Chile, United States (1960). Prince William Sound, United States (1964)
Volcanic events	Volcanic activities are confined to specific countries and locations therein, hence not of a concern to many countries. Active warning systems, such as the one maintained by the United States. Few geological surveys helps anticipating the occurrence of eruption well in time	Instead of recovery planning, the only strategy that can best be adopted is to avoid locating critical facilities in the vulnerable areas	The area of impact could be very large, which spreads not only to the immediate hazard area but also to areas to which the ash can spread out. Fine dust particulates and volcanic ash can cause flashover in insulators and potentially damage transformers	Indonesia is considered most vulnerable to volcanic eruptions followed by countries like Philippines, Japan, Mexico, Ethiopia, Guatemala, Ecuador, Italy, El, Salvador, and Kenya

Wildfire	Generally, no warning period is available	Facilities charred down may take days or weeks to get restored	Major power system components, substations, and transmission systems lying in the fire affected areas will only get damaged and hence the operators may need to divert the path of power flow avoiding the affected areas. Effective vegetation management can often help limit the damage caused by wildfire, although very large fires can often jump even the most aggressive protective margins	Going by the claims of the scientists working on climate change, more frequent and more intense wildfires are predicted in upcoming times

reliable. A resilient EPS should also be capable of predicting and preparing itself to face any eventualities with potential to cause power outage. A resilient EPS must also have the inherent mechanism to regroup itself to bounce back to normalcy immediately after the occurrence of disruptive events ensuring power supply to critical establishment at the least. A reliable power supply under any eventuality should be the hallmark of any resilient power system. And in this context, there happens the possibility of mistaking resiliency as reliability and the vice versa. However, it is not so though there is a strong resemblance between the two. Reliability is well defined and understood in the context of electricity sector and has well-defined metrics contrary to resiliency.

3.1 Understanding resiliency

The National Association of Regulatory Utility Commission (NARUC) (Andrews, 1995) defines resilience as—"the ability of the system to anticipate, absorb, recover from, and adapt to disruptive events, particularly high-impact, low-frequency events." The US Department of Homeland Security (DHS) (House, 2013) provides a definition of resilience as "the ability to adapt to changing conditions and withstand and rapidly recover from disruption due to emergencies." The National Infrastructure Advisory Council (NIAC) (Vugrin et al., 2010) defines resilience in terms of infrastructure resilience as—"the ability to reduce the magnitude and/or duration of disruptive events. The effectiveness of a resilient infrastructure or enterprise depends upon its ability to anticipate, absorb, adapt to, and/or rapidly recover from a potentially disruptive event." The National Association of Regulatory Utility Commission (NARUC) (Keogh & Cody, n.d.) in its report on "A Framework for Establishing Critical Infrastructure Resilience Goals," submitted to President Obama in November 2010, provided a broad-based definition of resilience: "Infrastructure resilience is the ability to reduce the magnitude and/or duration of disruptive events. The effectiveness of a resilient infrastructure or enterprise depends upon its ability to anticipate, absorb, adapt to and/or rapidly recover from a potentially disruptive event." allowing for sector-specific applicability. The framework was intended to narrow down the interpretation of resilience for various sectors in terms of their abilities under the qualitative parameters: (1) robustness, (2) resourcefulness, (3) rapid recovery, and (4) adaptability. The report also opines that resiliency as interpreted based on the above-mentioned criteria would help formulate policies in view of national security. However, the definition appears imprecise to be used in a regulatory framework. In this context, NARCU provides a rather more pragmatic definition of resilience as—*"the term 'resilience' means the ability to prepare for and adapt to changing conditions and withstand and recover rapidly from disruptions. Resilience includes the ability to withstand and recover from deliberate attacks, accidents, or naturally occurring threats or incidents."* Going by all these aforementioned definitions, it may be inferred that modern day infrastructure (the EPS in context of this study) should have few common traits incorporated to be resilient enough. They may be listed out as (S. Chanda & Srivastava, 2015a, 2015b): (i) ability to "withstand" any sudden inclement weather or human attack on the infrastructure, (ii) mechanism to "respond" quickly and restore continuity of service as soon as possible, after the occurrence of the unavoidable disruptive event, (iii) flexibility to "adapt" to then new evolving conditions after the event passes away, and (iv) the ability to "predict" or foresee the future disruptive events based on experiences gained from past occurrences.

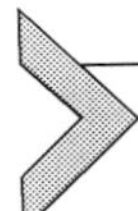

4. Resiliency versus reliability of electrical power systems

Resiliency of an EPS can be gauged by the occurrence of LFHI events in terms of the system's preparedness to deal with, in its dexterity of handling and restoring the critical loads in quickest possible time prioritizing over the noncritical loads. Reliability on the other hand relates to the system's ability to provide uninterrupted service overcoming the frequently happening HFLI events. The HFLI events are essentially the system disturbances which the protection schemes fail to suppress, such as short circuits due to animal intrusion, etc. causing very short duration outages. Reliability of power supply needs to be ensured for all level of consumers connected to the system. A brief comparison between resiliency and reliability is provided in Table 1.2 (S. Chanda & Srivastava, 2015b).

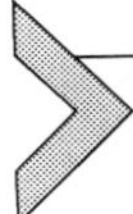

5. Matrices for measurement of resiliency and reliability

5.1 Reliability matrices

The degree of reliability required varies from grid to grid or nation to nation. What is sufficient for one nation may not be of acceptable level for the other. In this context, quantification of reliability through measurable matrices provides a clear understanding of the reliability requirement. Fortunately,

Table 1.2 Resiliency versus reliability.

Parameters for comparison	Resiliency	Reliability
Nature of disruptive event	LFHI	HFLI
Hindrance	Network design, operating conditions, control and demand side management (DMS) actions in response to disruptive events	The need to ensure uninterrupted quality power supply to all connected loads
Definition	Not well defined	Well defined
Matrices for measurement	Yet to be defined properly	SAIFI, SAIDI, ENS, CAIDI, CAIFI
Duration of measurement	Measured before or after an event	Usually measured over a certain period of time
Load priority	Focuses on recovery of critical loads followed by others	All connected loads carry equal priority
Types of outages needed to be dealt with	All outages irrespective of duration	Short duration outages less than 5 min are not accounted

matrices for reliability measurement are well defined. Reliability of a system is calculated in terms of duration and frequency of outages. For calculation of reliability in terms of duration of outages, the often used matrices are, namely, system average interruption duration index (SAIDI) and the customer average interruption duration index (CAIDI). Eqs. (1.1), (1.3) provide the formulation of SAIDI and CAIDI, respectively. Eq. (1.2) provides the formulation for reliability calculated using system average interruption frequency index (SAIFI), which is a measure of reliability based on frequency of outages.

$$\text{SAIDI} = \frac{\text{Sum of duration outages suffered by customers}}{\text{Total numbers of customers serviced}} = \frac{\sum_i d_i n_i}{n_t} \tag{1.1}$$

$$\text{SAIFI} = \frac{\text{Total number of outages suffered by customers}}{\text{Total numbers of customers serviced}} = \frac{\sum_i n_i}{n_t} \tag{1.2}$$

$$\text{CAIDI} = \frac{\text{Sum of duration of outages suffered by customers}}{\text{Total numbers of outages suffered by customers}} = \frac{\sum_i d_i n_i}{\sum_i n_i} = \frac{\text{SAIDI}}{\text{SAIFI}} \tag{1.3}$$

There are few more formulas to measure reliability, such as customer average interruption frequency index (CAIFI) and momentary average interruption frequency index (MAIFI), which are as listed in Table 1.3. Interruption cost estimate (ICE) calculator, developed by U.S. Department of Energy (DOE), is another metric which helps estimating interruption costs and benefits on investing in infrastructure development to enhance reliability.

In the context of reliability matrices, the demarcating points exposing the difference between reliability and resilience can be deliberated as follows. All the available reliability matrices are formulated based on frequency and duration of events, an apparent indication to HFLI events. When the frequency of occurrence of events tends to be minimal and duration of outages tends to be relatively of infinite time, which are the attributes of LFHI events, these aforementioned reliability matrices become redundant. In addition, matrices like ICE calculator, which again is devised for short duration events, do not take into consideration the fact that the cost of lost power keeps increasing with duration of outage. In short, the reliability matrices cannot be extended to serve as a measure of resilience of a system.

5.2 Resiliency matrices

The very nature of LFHI events makes resilience subjective as it becomes next to impossible to quantify resiliency of a system. Each type of LFHI event has different levels of unpredictability and impact which makes the preparedness to face all types of eventualities almost impossible. A well-defined quantifier of resiliency in this context would certainly help the planners and operators to decide upon the required degree of preparedness. Preparedness requires large investment and the benefit of the investment needs to be known prior to make appropriate tradeoff between investment and degree of resiliency.

There have been several attempts at development of resilience matrices based on varied approaches. As threats to resilience and their consequences do not have common attributes, a generic resilience metrics is not feasible

Table 1.3 Reliability matrices.

Reliability matrices	Formulation	Purpose of the metric
SAIDI (system average interruption duration index)	$\frac{\sum_i (r_i n_i)}{n_t}$	(1) Most used measure for sustained interruption. (2) Measures the total duration of an interruption for the average customer during a given time period. (3) Calculation made on monthly or yearly basis
SAIFI (system average interruption frequency index)	$\frac{\sum_i n_i}{n_t}$	Indicates the average count of outages faced by the customer in a given year (or time period as may be specified for a particular study)
CAIDI (customer average interruption duration index)	$\frac{\sum_i (r_i n_i)}{\sum_i n_i}$	Measures the average time to restore service
CAIFI (customer average interruption frequency index)	$\frac{\sum_o (n_o)}{\sum_i (n_i)}$	Measures the average count of interruptions per customer interrupted per year
CIII (customer interrupted per interruption index)	$\frac{\sum_i (n_i)}{\sum_o (n_o)}$	Provides the average counts of customers facing interruption as a result of an outage
MAIFI (momentary average interruption frequency index)	$\frac{\sum_i (Id_i n_i)}{n_t}$	Measures the average number of small-time interruptions (mostly that occurs at the substation, as the momentary interruptions at in distribution systems are difficult to trace) over a given fixed period of time
ASAI (average service availability index)	$\left[1 - \left(\frac{\sum_i (r_i n_i)}{n_t}\right)\right] \times 100$	(1) It is a measure of hours of service available to customers as against hours of services demanded over a given time period. (2) It generally indicates to the reliability of service

Id_i = number of interrupting device operations; n_i = total count of customers interrupted; n_o = count of interruptions; n_t = total count of customers served; r_i = restoration duration; T = time period under study.

and researchers have proposed various approaches to deal with this issue. S. Chanda (2015) opines that the impact of any disruptive event can best be assessed by the time taken by the system to bounce back to normalcy. Some researchers have deliberated upon formulation of resiliency metrics encompassing the entire infrastructure of a community or a city (Fisher & Norman, 2010; Petit et al., 2013). Sandia National Laboratories on the other hand have proposed a "Resilience Analysis Process" to assess and evaluate the resiliency of a given system for further improvement in addition to helping the policy makers reach to a dependable, risk-based decision (Watson et al., 2014). The first six steps of the RAP provide the means to assess a system's baseline performance. As all seven steps are practiced, RAP scopes into identifying means of improvements to achieve increased resilience. A graphic representation of the seven-step resilience assessment process is enumerated in Table 1.4. The framework also envisions that metrics designed for certain set of threats should reflect the system performance, as not only the system resilience should be capable of dealing with a certain threat but also it should be able to do so with certain desired level of performance. In addition, resilience matrices should have certain helpful attributes, as listed below.

Easy comprehension and interpretation: Operators should be able to understand the situation and take appropriate action in the quickest possible time.

Flexible and scalable: To incorporate resilience related measures with minimal modifications.

Least metric computation time: Metric computation time should be less than system control response time.

Metric sensitivity: The sensitivity of the metric should corroborate to physical changes in the network.

Post event capturing of attributes: The metric should be able to capture all the attributes of the impact of the disruptive events on the power system, specifically quality and continuity of service. In other words, the metric should preserve maximum information about all the noncommensurate factors that affect power system resilience.

Easy implementation: The metric should be easily implementable in the distribution management systems (DMS) and compatible with existing and future data acquisition hardware.

Metric format for data exchange: The format for resilience metric data exchange should be following the common data exchange protocols.

Table 1.4 Resilience analysis process as proposed by Sandia National Laboratory.

Sequence of analysis	Process sequence title	Process sequence objectives/exercises
Step-1	Define resilience goals	(1) Fixing of high-level resilience goals. (2) Define the kind and magnitude of changes to be incorporated. (3) Identify key stakeholders and anticipate possible goals of conflicting interest among them
Step-2	Set power system and its resilience metrics	(1) Scope of the analysis established in terms of system's geographic boundaries, relevant time periods, and/or relevant components. (2) Matrices selection to make operational and planning decisions
Step-3	Characterization of threats	(1) To estimate the level of resilience needs to be obtained against identified threat(s) through characterization of the threat(s). (2) Characterization of the threat and its consequences to figure out the most important aspects of the threat that needs to be addressed to minimize the impact
Step-4	Assessment of the anticipated level of impact	Picking up the attributes of each threat to assess the extent of damage that is likely to result.
Step-5	Development and implementation of system models	(1) Input from Step-4 to develop system models—making association of damage to output level of the designed system models. (2) As many system models as may be required needs to be developed to capture information on relevant aspects of the whole system. (3) Presence of dependencies between these models needs to be identified
Step-6	Calculate consequence	Outputs from system models are converted to the resilience metrics that were defined during Step-2
Step-7	Evaluate resilience improvements	(1) Decisions to be made about modification in operational decisions and/or plan investments to improve resilience. (2) The newly populated metrics for resilience are used to decide upon need for a physical change/a policy change/a procedural change

6. Means of improving resiliency

The present-day electrical power system (EPS), which is fast evolving as a smart grid, the microgrid, is viewed as an important enabling component. Although implementation of microgrid technology in the existing traditional EPS is associated with many issues and challenges, it also brings along immense benefits in smart grid context. One of the major benefits could be cited as the role of microgrid in improving the system resiliency (Patnaik et al., 2020). In case of loss of mains, microgrid ensures continuity of supply to its loads with the help of distributed generators (DGs) and energy storing systems (ESSs) provisioned within its defined electrical boundary. In case of LFHI or any other disruptive events, the microgrids can ensure continuity of supply during or immediately after the occurrence of the events, thereby contributing to the resilience of the power distribution system. Improving power system resiliency thus can be viewed as an important complementary value proposition of microgrids. However, during the happening of a high-impact disruptive event, the very survival of microgrids operating in islanded mode and serving locally connected load is itself a crucial issue in the context of system resiliency. In the above context, Balasubramaniam et al. (2016) have proposed an energy management system in microgrids operating in islanded mode because of disruption of supply from the main grid for a longer period of time. The loads in the microgrid are characterized as critical and noncritical loads. The task of prioritizing loads for internal resourcing with an objective of maintaining the service continuity alive till the main supply is restored is viewed as a nonlinear programming problem. In similar context, the study by Khodaei (2014) proposes resiliency-oriented microgrid optimal scheduling model, wherein the load shedding within the microgrid is minimized through efficient scheduling of available resources when supply from main utility grid gets disrupted for a prolonged period of time. Quick and efficient system restoration can be realized through the increasing number of DGs being penetrated to the EPS. In Jiang et al. (2018), the authors have investigated the effect of DG's switching-in points over areas of restoration islands in a distribution network. The study also introduces a discrete particle swam optimization algorithm-based islanding scheme with an aim to restore maximum number of loads aftermath of extremely disruptive events.

Though restoration of power supply to residential customers does not come first in the priority list, a resilient system must address this at the earliest to save people from undue inconvenience. Rahimi and Chowdhury (2014) have addressed this issue by proposing an approach based on combinatorial use of plug-in hybrid electric vehicle (PHEV) and a photovoltaic (PV) resource to enhance customer serviceability, thereby enhancing the resiliency of the distribution system.

Modern day grid control is highly interactive and hence has heavy dependence on robust communication systems. Wide-area measurement systems provide the necessary control inputs in the form of synchrophasor data, which is very vital for wide-area damping control (WADC) used for damping out interarea oscillations caused by increased penetration of renewable resources and resulting uncertainty. Communication systems always remain highly vulnerable to extreme disruptive events (such as cyberattacks) and failure of it will certainly affect the control system of the power system resulting in system instability. This aspect of system resiliency issues is addressed by a few studies as follows. S. Zhang and Vittal (2013, 2014) have proposed a hierarchical framework based on a set of wide-area measurements for control. It involves switching of channels based on mathematical morphology identification to nullify the impact of communication failures on control effectiveness. The control framework is designed consisting of a number of single-input single-output (SISO) supplementary damping controllers associated with a static VAR compensator. When a signal in the hierarchy fails because of a communication failure, the control automatically identifies another available signal in an alternate communication route ensuring the stability of the system. It leverages the large investment in nationwide installation of phasor measurement units. In similar context, Bento et al. (2018) have proposed a WADC design method focused on the need for robustness at various operating points of the power system, to deal with the delay in the communication channels, and to tackle possible situation of permanent loss of communication signals for effective functioning of important controllers. Some extremely disruptive events (e.g., Hurricanes) and their path of progression can be foreseen and monitored well in advance so as to put in place preventive control actions. The distribution phasor measurement units (D-PMUs) come handy in such cases, and synchrophasor preevent reconfiguration can be planned for enhanced resiliency in power distribution systems. However, the D-PMUs is also vulnerable to the disruptive effects of events like cyberattacks. In this context, Pandey et al. (2020) have presented a data mining-based approach for detection of anomaly in DPMUs and a resiliency-oriented preevent reconfiguration with islanding as proactive mechanisms to minimize the impact of adverse events on system using processed synchrophasor data.

Increased presence of microgrids, distributed generators, and distributed resources (such as behind-the-meter energy storage systems and electric vehicles) in present-day EPSs result in large number of control points staggered over the nationwide utility grid. A centralized control architecture though is very much advantageous and becomes a major cause of hindrance itself during the occurrence of an extreme event. The communication bandwidth, latency, and the scalability of centralized control architecture limit the ability of the EPS to use the distributed resources and devices as active resources to survive and recover from an extreme event. Decentralized or distributed control architecture rather serves well in providing the EPS with higher resiliency against extreme events. In this context, a centralized control architecture during normal circumstances and a coordinated central and distributed control architecture to face extreme events for enhanced resiliency is proposed by Colson et al. (2011) and Schneider et al. (2019). Increased operational flexibility obtained through coordination between central and distributed control systems is envisioned in Schneider et al. (2019), whereas a decentralized multiagent control method for distributed microgrids is introduced in Colson et al. (2011) which deliberates upon a dispersed decision-making approach engaging smart microgrid control agents who act collaboratively during normal and emergency situations. Such agent-based control in conjunction with microgrid technology can improve power system resiliency significantly and is viewed as an enabling technology for future smart grids.

A framework for resiliency-oriented proactive recovery of electric power assets against impact of hurricane is suggested in A. Arab, Khodaei, Han, and Khator (2015). The framework envisages designing the next generation decision-making tool for proactive recovery through several coordinated models: (1) the outage model that assesses the impact of the event on power system components; (2) the crew mobilization model that manages the resources before the happening of the event; and (3) the posthurricane recovery model that helps manage resources post occurrence of the event. It is envisaged to further extend these models to establish applicability of them to different kinds of electric grids with different technologies and regulatory issues. A. Arab, Khodaei, Khator, et al. (2015) have similarly proposed a proactive resource allocation model for repair and restoration of power system infrastructure located on the path of an impending hurricane. The problem is modeled as a stochastic integer program with complete recourse. The large-scale mixed-integer equivalence of the original model is solved by the Benders' decomposition method to handle computation burden. Intentional islanding can be used as an effective means to arrest the cascading outages ensued by disruptive events. However, this requires strategy to position and manage the microgrids to optimally control their internal resources. In this respect, Gholami et al. (2017) have proposed a two-stage adaptive robust formulation for predisturbance scheduling of microgrid. The approach consists of obtaining the best day-ahead schedule prepared considering different uncertainties followed by a decomposition strategy, called as column-and-constraint generation (C&CG) algorithm, to realize the above objective. A similar approach for enhanced system resilience improvement strategy is proposed in Huo and Cotilla-Sanchez (2018), which suggests adequate partitioning of the entire power grid and positioning different areas of the grid at good starting points for survival and reconnection. The method is based upon a multiobjective electrical distance-based clustering algorithm with an index that ensures adequate amount of generation capacity for each cluster, an electrical distance matrix based on Q-V sensitivity, and a simplification of the clustering method that shortens simulation times by equivalencing a given power system model.

7. Conclusion

The importance of resiliency as a key factor in present business models has gained much attention as increasing instances of outages because of extreme climatic conditions have burnt a hole in the balance sheets of energy centric businesses. The loss caused by outages has outweighed the benefits accrued by minimizing the energy unit price. Hence, investment in technology upgradation and infrastructure augmentation with an objective of improving power system resilience is a subject matter of growing importance and requires researchers' attention to this Greenfield research area. Ensuring uninterrupted power supply to all the customers connected to the utility by mitigating and/or overriding frequently occurring temporary as well as permanent system faults is what preserving the reliability of the system is all about. A resilient system on the other hand should fight out long duration outages caused by low-frequency high-impact (LFHI) events; it must be able to predict, prepare for, endure, and bounce back to normalcy before, during, and after the happening of a severely disruptive event.

As these so-called LFHI events are rare and unpredictable, a resilient system must also learn from its past experiences and adapt itself to be resilient against the re-occurrence of such events. And this requires indices or matrices to capture the attributes of an event to set a desired resilient benchmark against specific or a group or all possible extreme events. There are many established and proven matrices to evaluate reliability of a system. But generalized indices to measure and evaluate the resiliency of the modern power systems against varied kinds of extreme events is yet to be put in practice. As has been discussed in previous sections, there have been many studies proposing methods and schemes to enhance the resilience of a utility grid. Some of these are based on conventional grids, making them obsolete in context of the evolving smart grids. Some suggest making the infrastructure robust against possible physical damages. Whereas making the communication system secured against the cyberattacks is another means of adding to the resilience of a system, and various cyber security systems in this context have been proposed by many. Although there has been several different approaches to enhance the power system resiliency prescribed in literature, a more generalized resilience evaluation, quantification, and economic realization is still awaited and leaves ample scope of study for future researchers.

References

Andrews, C. J. (1995). Evaluating risk management strategies in resource planning. *IEEE Transactions on Power Systems*, *10*(1), 420–426. https://doi.org/10.1109/59.373966.

Arab, A., Khodaei, A., Han, Z., & Khator, S. K. (2015). Proactive recovery of electric power assets for resiliency enhancement. *IEEE Access*, *3*, 99–109. https://doi.org/10.1109/ACCESS.2015.2404215.

Arab, A., Khodaei, A., Khator, S. K., Ding, K., Emesih, V. A., & Han, Z. (2015). Stochastic pre-hurricane restoration planning for electric power systems infrastructure. *IEEE Transactions on Smart Grid*, *6*(2), 1046–1054. https://doi.org/10.1109/TSG.2015.2388736.

Balasubramaniam, K., Saraf, P., Hadidi, R., & Makram, E. B. (2016). Energy management system for enhanced resiliency of microgrids during islanded operation. *Electric Power Systems Research*, *137*, 133–141. https://doi.org/10.1016/j.epsr.2016.04.006.

Bento, M. E. C., Kuiava, R., & Ramos, R. A. (2018). Design of wide-area damping controllers incorporating resiliency to permanent failure of remote communication links. *Journal of Control, Automation and Electrical Systems*, *29*(5), 541–550. https://doi.org/10.1007/s40313-018-0398-3.

Chanda, S. (2015). *Easuring and enabling resiliency in distribution systems with multiple microgrids*. Doctoral dissertation Washington State University.

Chanda, S., & Srivastava, A. K. (2015a). Quantifying resiliency of smart power distribution systems with distributed energy resources. In *2015 IEEE 24th international symposium on industrial electronics (ISIE)* (pp. 766–771). https://doi.org/10.1109/ISIE.2015.7281565.

Chanda, S., & Srivastava, A. K. (2015b). Defining and enabling resiliency of electric distribution systems with multiple microgrids. *IEEE Transactions on Smart Grid*, 7(6), 2859–2868. https://doi.org/10.1109/TSG.2016.2561303.

Colson, C. M., Nehrir, M. H., & Gunderson, R. W. (2011). Distributed multi-agent microgrids: A decentralized approach to resilient power system self-healing. In *Proceedings—ISRCS 2011: 4th international symposium on resilient control systems* (pp. 83–88). https://doi.org/10.1109/ISRCS.2011.6016094.

Enhancing the Resilience of the Nation Electricity System. (2017). http://nap.edu/24836.

Fisher, R. E., & Norman, M. (2010). Developing measurement indices to enhance protection and resilience of critical infrastructure and key resources. *Journal of Business Continuity & Emergency Planning*, *4*(3), 191–206.

Frank, T. (2019). *E&E News reporter*. https://www.eenews.net/stories/1061245945.

Gholami, A., Shekari, T., & Grijalva, S. (2017). Proactive management of microgrids for resiliency enhancement: An adaptive robust approach. *IEEE Transactions on Sustainable Energy*, *10*(1), 470–480. https://doi.org/10.1109/TSTE.2017.2740433.

House, W. (2013). Critical infrastructure security and resilience. *Vol. 12. Presidential Policy Directive/PPD–21*. US: White House. https://www.govinfo.gov/content/pkg/PPP-2013-book1/pdf/PPP-2013-book1-doc-pg106.pdf. (Accessed 5 January 2022).

Huo, C., & Cotilla-Sanchez, E. (2018). A power-balanced clustering algorithm to improve electrical infrastructure resiliency. In *20th power systems computation conference, PSCC 2018* Institute of Electrical and Electronics Engineers Inc. https://doi.org/10.23919/PSCC.2018.8442565.

Hussain, A., & Pande, P. (2020). https://www.bloomenergy.com/blog/2020-predictions-top-energy-trends-were-anticipating-this-year.

Jiang, Z., Wang, L., & Chen, W. (2018). A distribution network islanding strategy based on topological characteristics to improve system resiliency. In *2nd IEEE conference on energy internet and energy system integration (EI2), Beijing, Beijing*. https://doi.org/10.1109/EI2.2018.8582086.

Keogh, M., & Cody, C. (n.d.). Resilience in regulated utilities. National Association of regulatory utility commissioners. http://www.naruc.org/Grants/Documents/Resilience%20in%20Regulated%20Utilities%20ONLINE%2011_12.pdf.

Khodaei, A. (2014). Resiliency-oriented microgrid optimal scheduling. *IEEE Transactions on Smart Grid*, *5*(4), 1584–1591. https://doi.org/10.1109/TSG.2014.2311465.

Pandey, S., Chanda, S., Srivastava, A. K., & Hovsapian, R. O. (2020). Resiliency-driven proactive distribution system reconfiguration with synchrophasor data. *IEEE Transactions on Power Systems*, *35*(4), 2748–2758. https://doi.org/10.1109/TPWRS.2020.2968611.

Patnaik, B., Mishra, M., Bansal, R. C., & Jena, R. K. (2020). AC microgrid protection—A review: Current and future prospective. *Applied Energy*, 271. https://doi.org/10.1016/j.apenergy.2020.115210.

Petit, F. D., Fisher, R. E., & Veselka, S. N. (2013). *Resilience measurement index: An indicator of critical infrastructure resilience*. Argonne National Laboratory. https://publications.anl.gov/anlpubs/2013/07/76797.pdf.

Rahimi, K., & Chowdhury, B. (2014). A hybrid approach to improve the resiliency of the power distribution system. In *2014 North American power symposium (NAPS), Pullman, WA, USA. IEEE*. https://doi.org/10.1109/NAPS.2014.6965472.

Schneider, K. P., Laval, S., Hansen, J., Melton, R. B., Ponder, L., Fox, L., Hart, J., Hambrick, J., Buckner, M., Baggu, M., Prabakar, K., Manjrekar, M., Essakiappan, S., Tolbert, L. M., Liu, Y., Dong, J., Zhu, L., Smallwood, A., Jayantilal, A., … Yuan, G. (2019). A distributed power system control architecture for improved distribution system resiliency. *IEEE Access*, 7, 9957–9970. https://doi.org/10.1109/ACCESS.2019.2891368.

Vugrin, E. D., Warren, D. E., Ehlen, M. A., & Camphouse, R. C. (2010). A framework for assessing the resilience of infrastructure and economic systems. In *Sustainable and resilient critical infrastructure systems* (pp. 77–116). Springer. https://doi.org/10.1007/978-3-642-11405-2_3.

Watson, J. P., Guttromson, R., Silva-Monroy, C., Jeffers, R., Jones, K., Ellison, J., Rath, C., Gearhart, J., Jones, D., Corbet, T., & Hanley, C. (2014). *Conceptual framework for developing resilience metrics for the electricity oil and gas sectors in the United States*. Sandia National Laboratories. https://www.energy.gov/sites/prod/files/2015/09/f26/EnergyResilienceReport_Final_SAND2014-18019.pdf.

Zhang, S., & Vittal, V. (2013). Improving grid resiliency using hierarchical wide area measurements. In *IEEE power and energy society general meeting*. https://doi.org/10.1109/PESMG.2013.6672307.

Zhang, S., & Vittal, V. (2014). Wide-area control resiliency using redundant communication paths. *IEEE Transactions on Power Systems*, *29*(5), 2189–2199. https://doi.org/10.1109/TPWRS.2014.2300502.

CHAPTER TWO

Resilient smart-grid system: Issues and challenges

Manohar Mishra[a], Bhaskar Patnaik[b], and Ramesh C. Bansal[c,d]

[a]Department of Electrical and Electronics Engineering, FET, Siksha "O" Anusandhan University, Bhubaneswar, India
[b]Biju Patnaik University of Technology, Rourkela, Odisha, India
[c]Department of Electrical Engineering, University of Sharjah, Sharjah, United Arab Emirates
[d]Department of Electrical, Electronic and Computer Engineering, University of Pretoria, Pretoria, South Africa

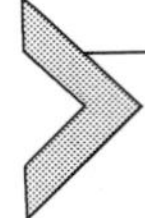

1. Introduction

1.1 Evolution of power system

The year 1891, when the first major electrical power system (EPS) was commissioned to drive a 100-hp (75 kW) synchronous electric motor, could well be considered as the beginning of flourish of the commercial application of electricity. Since then, the power system has gone through mammoth expansion ushering in rapid industrialization and giving a new fillip to the world economy. Over the past century, electrical energy has been the key trigger in rapid worldwide urbanization and infrastructural development and the same in turn caused a sharp increase in demand for power, leading to massive expansion of the EPS. The growth of the EPS has largely remained influenced by socioeconomic, political, and geographical constraints specific to the region or countries concerned (Farhangi, 2010). However, the basic topology of the existing EPS has primarily stayed unchanged. A strictly hierarchical system of unidirectional power flow from the generating station to the load and clearly demarcated operational framework of generation, transmission, and distribution mark the conventional EPS. Till a decade ago, the EPS or for that matter the electrical grid was characterized by (i) centralized generation, (ii) electro-mechanical components, (iii) one-way communication, (iv) manual monitoring and manual restoration assisted by few sensors, and (v) limited customer choice. In addition, the burgeoning size of the EPS made it humanly impossible to monitor and control the grid manually. Thus, the control was limited and hence grid failures and blackouts were regular features. Also, the design of the EPS was confined to the primary objective of withstanding the occasional maximum anticipated peak load demand and

https://doi.org/10.1016/B978-0-323-85536-5.00010-2

hence was not in a position to dynamically schedule load to have an efficient power management. The EPS thus was inherently inefficient and needed to evolve and smarten up to the newer requirements. Application of information and communication technology (ICT) to have effective monitoring and control of power through aggregation and analysis of the data obtained from sensing devices placed at various locations of the EPS was next thing to come as a precursor to a smarter grid system. Deployment of supervisory control and data acquisition (SCADA) in this sense can be cited as an instance of traditional grid taking a leap toward modernization.

1.2 Need for a smarter grid

At the advent of electricity, having access to it was considered a privilege, and stability, reliability, quality, and price of power supplied were never questioned and were considered trivial. However, with passing time and with increased dependency on electric power, the aforementioned factors became issues of prime importance. The consumers became more discerning on their rights to access for quality power. Nonetheless, advanced technologies have helped EPSs to imbibe better monitoring and control strategies toward realization of stability and reliability of service. Inculcation of cutting-edge technologies in the field of communication engineering and information technology into the physical power system has converted the traditional system into a smart system, namely, cyber-physical system (CPS). The CPS essentially comprises the existing EPS being controlled and monitored dynamically by instructions received by a central decision-making unit. The central unit makes its decision based on processing and analyzing of data received from numerous sensors placed at various locations of the EPS. So, advanced sensing devices, a robust communication channel, an efficient data acquisition system, and an intelligent fast-acting computing algorithm to make inference based on the available run-time information, make for the CPS.

Electricity became the key economic driver, and the availability of it at an affordable price has remained another aspect of concern right from day one. Power generation was and still is mostly fossil fuel based. There are two fundamental reasons that have made us to look for alternative renewable energy resources. First, increasing concern toward safety of the environment necessitates desisting from fossil fuel-based generation as it causes huge carbon emission. Second, and most importantly, the fossil fuel reserve is fast depleting, which not only accounts for increased price of electricity but also compels us to look for alternative energy sources. And this has lead to the search for renewable and sustainable energy resources. Over the last decade or so, there has been tremendous technological development in the field of renewable energy generation and leading to deployment of micro-sources, such as PV generators, micro-hydel power plants, and wind firms. These renewable sourced generators are located and integrated to the existing grid at various locations and voltage level in the distribution system of the grid. Named as distributed generators (DGs), these micro-sources are connected to the grid at lower voltage level, are of small capacity, and generally commissioned and operated by persons or entities that are originally the consumers connected to the main utility grid. This is where the concept of microgrid creeps in. The concept of microgrid and its role as a smart-grid enabler is discussed in the later section. However, introduction of DGs or for that matter the microgrids radically changes the topology and functioning of the EPS. The topology becomes dynamic, power flow is bidirectional, generation is decentralized, and the strictly hierarchical construct of the traditional EPS gets extinct.

1.3 Smart grid: The next generation electric power system

In present time, the power system network is behemoth, spanning the globe to meet the ever-increasing demand for electricity. However, it needs to get equipped to meet the demands of the 21st-century parameters such as quantity, quality, efficiency, reliability, ecology, and economy (Smart Grid Canada, 2012) (Federation, 2012). In this direction, great strides have already been made through inculcation of communication and information technology tools, making the earlier purely mechanical power system, a relatively self-controlled or self-organizing smart system. However, how smart a grid should be or for that matter what degree of modernization of the grid (power system and grid are two words used interchangeably meaning the same) would qualify a power system or grid to be designated as a smart grid is a question that demands a proper definition of a smart grid. It is interesting to know that smart grid as a concept is viewed differently by different consortiums and hence have different definitions. This is particularly so as what all are expected of a smart grid varies from perception to perception. Although grids equipped with smart meters were considered smart enough, the concept of smart grid has gone far beyond this as the demands have also changed radically. Power sector regulators in different countries have drawn up visions of modernizing their respective power system networks. They have also envisioned their perspective of a smart grid. A brief compilation of such viewpoints, definitions of smart grid, and expectations of a smart grid is listed out in Table 2.1, whereas Table 2.2 provides a

Table 2.1 Attributes of smart-grid and few pertinent definitions.

Consortiums	Attributes/benefits of smart-grid	Smart-grid definitions
EPRI (EPRI, 2021; Gellings, 2009)	• Allows optimization of the use of bulk generation and storage, transmission, distribution, distributed resources, and consumer end uses • Optimizes or minimize the use of energy • Mitigates environmental impact • Helps in managing assets and curtailing cost	"A smart grid is the use of sensors, communications, computational ability and control in some form to enhance the overall functionality of the electric power delivery system. A dumb system becomes smart by sensing, communicating, applying intelligence, exercising control and through feedback, continually adjusting"
US Department of Energy (DOE)'s vision (DOE, 2003, 2008; Ton, 2009)	• Ensures optimized use of assets and optimized operational efficiency • Seamless adoption of all types of generating units and energy storing devices • Ensures quality of power as is evinced for a digitally driven economy • Anticipates and responds to system disturbances in a self-healing manner • Operates resiliently against physical and cyberattacks and natural disasters • Facilitates dynamic consumers' participation • Enables new products, services, and markets	"The smart grid is the electricity delivery system, from point of generation to point of consumption, integrated with communications and information technology for enhanced grid operations, customer services, and environmental benefits." "A smart grid is self-healing, enables active participation of consumers, operates resiliently against attack and natural disasters, accommodates all generation and storage options, enables introduction of new products, services and markets, optimizes asset utilization and operates efficiently, provides power quality for the digital economy"

Table 2.1 Attributes of smart-grid and few pertinent definitions—cont'd

Consortiums	Attributes/benefits of smart-grid	Smart-grid definitions
ABB's vision (Jones, 2010)	• Adaptive in responding to fast changing operational conditions with reduced human interventions • Predictive, in terms of applying operational data to equipment maintenance practices and even identifying potential outages before they occur • Facilitates dynamic integration of communication and control features • Facilitates interaction between buyers and vendors • Optimized to maximize reliability, availability, efficiency, and economic performance • Secures from attack and naturally occurring disruptions	ABB's view on smart-grid veers beyond the inclusion of IT and smart meters. It emphasizes that smart-grid is an issue of political significance and involves many stakeholders who need to be kept appropriately informed. • Smart-grid is the result of continual evolution of the entire power network • Smart-grid includes T&D, focused on amalgamation of RERs, reliability, and grid-efficiency • Smart-grid involves demand-driven response and sustains potentially new technologies such as massive integration of electric vehicles (EVs)
European Union (EU)'s vision (ETP, 2006)	EU sees the smart grid as an active network: • To surmount the limitations in growth of distributed generation and storage • To facilitate interoperability and ensure supply security • Ensuring accessibility for all stakeholders into an unrestricted market • To minimize the impaction due to environmental effects on production and provisioning of electricity	Smart Grid includes both automation/IT and controllable power devices in the whole value chain from production to consumption. "A smart grid is an electricity network that can intelligently integrate the actions of all users connected to it—generators, consumers, and those that do both—to efficiently deliver sustainable, economic,

Continued

Table 2.1 Attributes of smart-grid and few pertinent definitions—cont'd

Consortiums	Attributes/benefits of smart-grid	Smart-grid definitions
	• To serve as an enabler of demand-side participation • To keep engaged customers' attention	and secure electricity supplies"
Electricite de France (EDF)'s vision (Herter et al., 2011; Miller, 2008)	• Empowered customers to decide upon their energy need and use accordingly, not only to cut cost but also to help sustain a clean energy • Backs the availability of demand-side resources for sale in the wholesale energy markets	EDF defines it as integrating distributed energy resources with dispersed intelligence and advanced automation
Hydro Quebe's vision (Abbey, 2008; Miller, 2008)	• To realize a dynamic distribution network • To obtain enhanced grid reliability • To achieve higher energy efficiency of the grid facilities • To achieve capacity enhancement through expansion and integration of newer sources of renewable energy sources (RES) and distributed generations (DGs) • To have optimized investments (financial and other) in the context of long-term operation, maintenance, and security of supply • To empower customers to be able to optimize their own consumption and thereby curtail on energy bills	Hydro Quebec has emphasized that a smarter grid is a necessity, not a choice

Table 2.1 Attributes of smart-grid and few pertinent definitions—cont'd

Consortiums	Attributes/benefits of smart-grid	Smart-grid definitions
General Electric's vision (Miller, 2008)		General Electric (GE) sees the smart grid "as a family of network control systems and asset management tools, empowered by sensors, communication pathways and information tools"
IESO's vision Ontario Independent Electricity System Operator (IESO)		Smart-grids involve many features, however in general, it veers around the application of IT tools that enhance the capabilities of the electrical power systems (EPSs) to deliver greater benefits to the customers through enhanced reliability, efficiency, sustainability, and customer control (through smart meters)
Ofgem's vision	Ofgem and its associated groups envision smart-grid as a smart electricity network that deploys smart communication networks, innovative technological products and services along with application of intelligent monitoring and control technologies to: • Enable adoption of generating units of all types, sizes, and technologies • Enable the demand side to help optimize system operation	

Continued

Table 2.1 Attributes of smart-grid and few pertinent definitions—cont'd

Consortiums	Attributes/benefits of smart-grid	Smart-grid definitions
	• Make available system balancing at the level of distribution systems and households • Empower customers to be highly informed and be able to make choice of supply • Reduce impaction of EPSs due to detrimental effects of environment • Ensure deliverance of appropriate degrees of reliability, flexibility, power quality, and security of supply	
OECD's vision (OECD, 2009)	Organization for Economic Cooperation and Development (OECD) characterizes smart-grid in two different perspectives: I. **Solution perspective: Smart-grid is**: • Energy efficient, through optimized energy usage, doing away with unnecessary capacity expansion. It ensured high power quality and security • Employs better energy monitoring and grid control strategies • Ensures better outage management through better data capture facilities • Has two-way flow of energy and real-time information, facilitating induction of green energy sources	

Table 2.1 Attributes of smart-grid and few pertinent definitions—cont'd

Consortiums	Attributes/benefits of smart-grid	Smart-grid definitions
	I. **Technical perspective: The smart-grid is**: A heavily intricate network of multiple digital and analog components and systems. The major components of a smart-grid are: (i) State-of-the-art grid components, (ii) Smart devices and metering systems, (iii) Smart communication systems, (iv) computational algorithms assisting in decision-making and human interfacing, (v) sophisticated control systems for easier facilitation of demand-side management and dynamic market functioning, and (vi) highly automated, responsive, and self-healing EPS involving seamlessly interfaced grid components	

Table 2.2 Smart-grid vs conventional grid (Chandel et al., 2015).

Conventional grid	Smart-grid
Electromechanical, solid state	Fully digital/microprocessor based
One-way and local two-way communication	Global integrated two-way communication
Centralized generation	Distributed generation
Limited monitoring, protection, and control system	Adaptive protection
Blind	Self-monitoring
Manual restoration	Automated restoration
Check equipment manually	Monitor equipment remotely
Consumers are uninformed and nonparticipative	Motivates and includes the consumer
Minimum optimization	Optimizes assets and operates efficiently

snapshot of how different a smart grid could be as against a conventional grid. It can be inferred from the compilation that the world is preparing and remodeling its power systems to meet the 21st-century requirements primarily through integration of communication systems, intelligent computing techniques, advanced metering infrastructures, advanced measuring and sensing devices, SCR-based control and protection devices, integration of distributed generating stations, etc. keeping in mind a greater level of consumer satisfaction through consumer participation, while keeping tab on the environmental health. In "Smart Grid System Report" July 2009, the US Department of Energy (2009) summarizes the distinctive characteristics of smart grid as:

- *Enables informed participation by customers*
- *Accommodates all generation and storage options*
- *Enables new products, services, and markets*
- *Provides the power quality for the range of needs*
- *Optimizes asset utilization and operating efficiency*
- *Operates resiliently to disturbances, attacks, and natural disasters*
- *Enables dynamic optimization of grid resources and operations*
- *Enables consumer participation and demand response*
- *Power disturbance savings*[a]

The progress toward smart grid as enumerated by Sandia National Laboratory envisions integration of renewable energy sources at Generation System Level, incorporation of wide-area monitoring and control at Transmission System Level, substation automation in distribution system level, and finally, advanced metering infrastructure along with EV/PHEV integration at consumer level, through integrated network of communication, information, and physical power infrastructure, as the way forward toward realization of a true smart grid.

Apart from the definitions of smart grid as listed in Table 2.1, a few recently coined ones with state-of-the-art technological developments, perception can be put as follows:

> *A smart grid is an electricity network enabling a two-way flow of electricity and data with digital communications technology enabling to detect, react and*

[a] The North American Electric Reliability Corporation, in its Reliability Performance Gap Index, indicates occurrences of 43 significant disturbances and outages in the United States in 2008, as against 30 such events in 2007. According to EAC report titled "Smart Grid, Enabler of the New Energy Economy," December 2008, Smart Grid technologies would reduce power disturbance costs to the US economy by $49 billion per year.

pro-act to changes in usage and multiple issues. Smart grids have self-healing capabilities and enable electricity customers to become active participants.

(Industry 4.0, 2021.)

Smart Grid is an Electrical Grid with Automation, Communication and IT systems that can monitor power flows from points of generation to points of consumption (even down to appliances level) and control the power flow or curtail the load to match generation in real time or near real time. Smart Grids can be achieved by implementing efficient transmission & distribution systems, system operations, consumer integration and renewable integration. Smart grids solutions help to monitor, measure and control power flows in real time that can contribute to identification of losses and thereby appropriate technical and managerial actions can be taken to arrest the losses.

There have been many pilot projects throughout the world to understand, assess, and analyze the challenges and benefits associated with deployment of smart grids. A few such cases are listed in Table 2.3. Moreover, Fig. 2.1 depicts the key driver, enabler, and challenges associate with smart grid.

Table 2.3 Few major smart grid pilot projects.

Smart-grid pilot projects/smart grid initiatives/reports on smart grid initiatives	Accrued/envisaged benefits	Nature of benefits
US Department of Energy Initiative (US Department of Electricity, 2012)	• Reduced peak load by 0.75–1.2 kW per customer	Techno—economic benefits
Queensland Smart City, Australia (US Department of Electricity, 2013)	• Reduction in peak load demand as well as consumption by 46% in the month of June 2012	
Smart Grid Pilot Project, Puducherry, India (Kappagantu et al., 2016)	• Peak saving of up to 66% leading to an average of 34.66% energy saving • 20% increase in revenue on replacement of 25% of the conventional meters with smart meters • Power quality improvement using APFS and active filters • Lesser interruptions, quick fault detection, and faster recovery of supply	

Continued

Table 2.3 Few major smart grid pilot projects—cont'd

Smart-grid pilot projects/smart grid initiatives/reports on smart grid initiatives	Accrued/envisaged benefits	Nature of benefits
Smart grid 2013 global report on 200 smart grid projects (US Department of Electricity, 2013)	• Increase in reliability to the tune of 9% witnessed in 70% of the pilot projects	
Study on ISO/wholesale market in US (EPRI, 2010)	• Reduction in peak by 10% (since 2006) due to demand-response (DR) contribution	
Duke Energy (Kappagantu & Daniel, 2018)	• Saving to the tune of $10.18 per consumer per annum in the city of Ohio, as reported by Duke Energy • Additional $3.5 of saving per consumer per annum on nonlabor expenses, such as meter testing/repairing/replacement	
Smart Grid Consumer Collaborative (SGCC) (Smart Energy Consumer Collaborative, 2013)	• SGCC claims to find a benefit of $2.00–$19.98 per consumer per annum through time of usage tariff	
Boulder, Colorado (Smart Energy Consumer Collaborative, 2013)	• SGCC reports that the PQ-related complaints by the consumers of Boulder, Colorado, have reduced to zero, from an average of 30, after adoption of SG	Benefit in terms of consumers' satisfaction
Smart Grid project in North America (J. Wang et al., 2011)	• Providing the service of web portals through which consumers can monitor their usage has led to enhanced consumers' appreciation for SG	
Smart Grid Consumer Collaborative (SGCC) (Smart Energy Consumer Collaborative, 2013)	• SGCC claims a reduction of 372 pounds of CO_2 emission per consumer through Volt/VAr control using US environmental protection agencies' estimate on "CO_2 equivalent estimate per kWh"	Environmental benefit

Table 2.3 Few major smart grid pilot projects—cont'd

Smart-grid pilot projects/smart grid initiatives/reports on smart grid initiatives	Accrued/envisaged benefits	Nature of benefits
	• Further an estimated reduction of 11–110 pounds per consumer per annum of CO_2 emission from time-varying rates using the same kWh equivalent (Smart Energy Consumer Collaborative, 2013), as SGCC reports	
Smart City of Queensland, Australia (US Department of Electricity, 2013)	• Reduction in greenhouse gases to the level of 54,000 tons in Queensland smart-city (Australia) by June 2012	

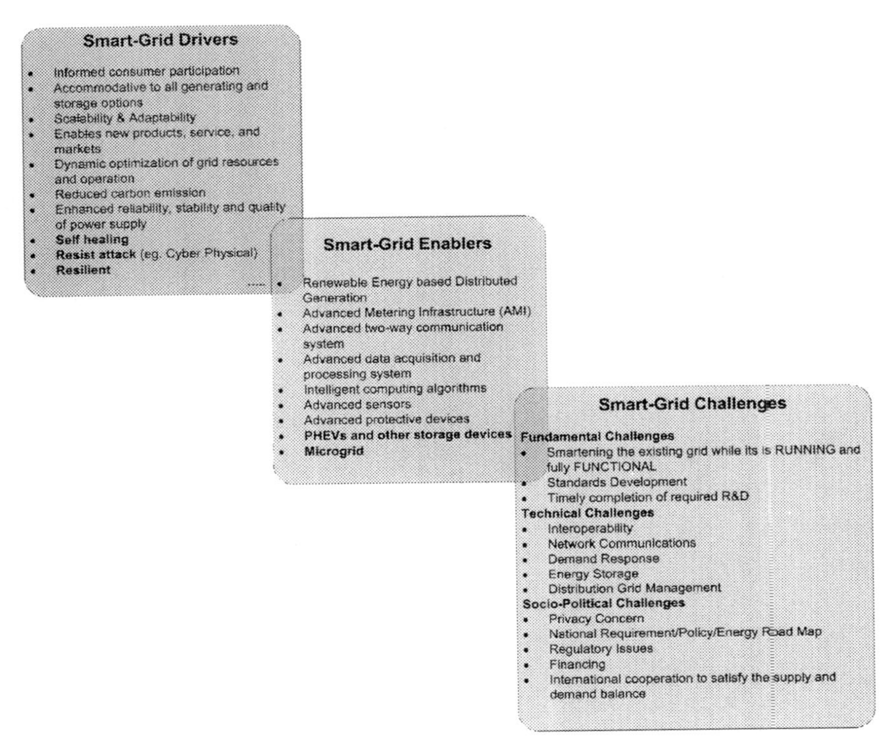

Fig. 2.1 Smart-grid key driver, enabler, and challenges. *(No permission required.)*

1.4 Microgrid: The key smart-grid enabler

Microgrid, considered as the smart-grid enabler, forms one of the major components of the smart grid. It contains all the components of the larger utility grid. Although smart grids are large-scale happenings at larger utility level, microgrids operate at smaller scale and can operate independent of the main grid: operating in islanded mode (Patnaik et al., 2020). Microgrids serve as the means to integrate distributed energy resources (DERs) to the low voltage (LV) networks at customers' end and in the process elevating the customer to be an active participant. Thus, microgrids as a subset of main grid offers various benefits, which includes increased system efficiency, reduced capacity expansion cost, improved power quality, and added system reliability.

A microgrid is basically a power island that facilitates exchange of power with the main utility grid during grid-connected mode. The microgrid while meeting the requirements of its locally connected loads, the excess of produced power is delivered back to the main grid. In case of reduced internal production, the microgrid receives the deficit power from the main grid. This need-based exchange of power between the microgrid and the utility not only calls for real-time load management but also makes the power flow within the Smart-grid (SG) bidirectional in nature. In this context, it is noteworthy that the traditional power grid has a centralized generation, long-distance meshed transmission lines, and radial distribution network. The SG on the other hand is a distributed generation, network of power islands meeting the local power load, with bi-directional power flow. Intentional islanding or need-based islanding can come as handy in keeping service continuity to critical infrastructures and thereby supplementing to the resiliency of a smart grid.

1.5 Resiliency: A new dimension in grid modernization

Although quantity, quality, efficiency, reliability, ecology, and economy are the key parameters that have been the driving factor in the worldwide effort for grid modernization and development of smart grid, the recent spurt in number of power outages due to climate change-induced extreme weather events and other such low-frequency high-impact (LFHI) disasters (which includes man made events as well) has raised the demand for a "resilient" power system. "Resiliency" has become the most sought after attribute of the 21st-century power system, as it has been felt that power outages and blackouts caused due to natural calamities have got serious repercussions on the economy of not only the stake-holding entities but also the entire nation. A detailed deliberation on "power system resiliency" can be found in Chapter 1. However, in a broader sense, power system resiliency can be summed up by the definition as given in M. Panteli et al. (2017) as:

> *the ability of a system to anticipate and withstand external shocks, bounce back to its pre-shock state as quickly as possible, and adapt to be better prepared to future catastrophic events.*

A system which is designed to effectively handle frequently occurring disruptive events (such as SC faults) is known to have high reliability. In contrast, the system is termed resilient when it is designed to remain in preparedness of a rarely occurring event of massive infrastructure damage capacity, withstands such shock by ensuring supply of power to the most critical loads, and bounces back quickly to the normalcy of the preevent state. Although reliability is a well-established and well-studied aspect of electrical power system, resiliency lacks with a generalized definition, quantifying matrices, and standards, which are so essential for resilience enhancement. When the conventional power system is transiting toward a smart system, understanding and implementing resilience enhancement measures is crucial. In this context, the subsequent chapters dwell upon the present status, challenges, and way ahead toward realization of a resilient smart-grid system.

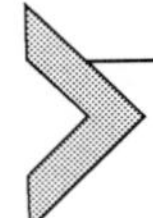

2. Resilient smart-grid system

2.1 Issues and challenges in smart grid deployment

The smart-grid, inclusive of the microgrid as a major component of its topology, enables energy management through two-way flows of electricity and information (Haravi & Ghafurian, 2011). The smart-grid architecture as depicted in Fig. 2.2 is a strong interface of networks, such as network of communication facilities interfaced with the network of power system physical infrastructure, leading to the coinage of the term "Cyber Physical System (CPS)." The network of bidirectional communication system is interfaced all along the power system structure segments of generation, transmission, end users, and also with all kinds of energy sources connected to the system.

Fig. 2.3 (Kabalci & Kabalci, 2019) is depictive of the bidirectional communication system which serves as the cyber layer over the power system physical layer. Embodiment of the necessary infrastructure and technology into the existing conventional grid, while it is running and fully functional is considered as a

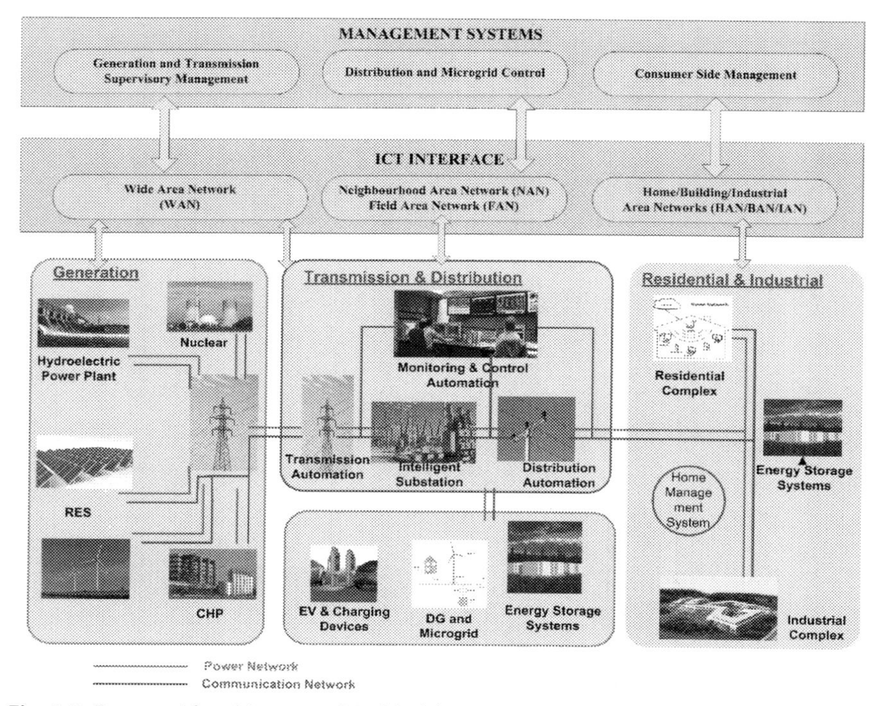

Fig. 2.2 Smart-grid architecture. *(Modified from Kabalci, E. & Kabalci, Y. (2019). Introduction to smart grid architecture. In Kabalci, E. & Kabalci, Y. (Eds.), Smart grids and their communication systems. Springer. https://doi.org/10.1007/978-981-13-1768-2_1.)*

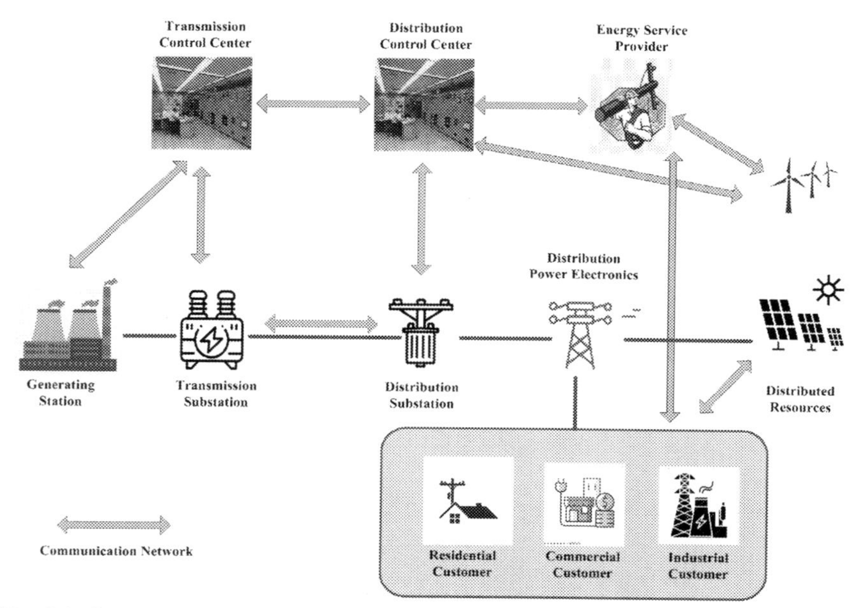

Fig. 2.3 Communication network in a smart-grid. *(Modified from Torres, J. (2011). Smart grid: Challenges and opportunities. https://www.osti.gov/servlets/purl/1120614.)*

fundamental challenge in smart-grid implementation. In addition, smart grid, which is a heterogeneous network of networks, would require efficient interoperability between networks, devices, sensors, and all other critical components therein to be able to perform to its desired objective.

Interoperability refers to "the ability of diverse systems and organizations to work together (interoperate). In the context of the electricity system, interoperability refers to the seamless, end-to-end connectivity of hardware and software from end-use devices through the T&D system to the power source, enhancing the coordination of energy flows with real-time information and analysis" (Gilbert et al., 2011). Interoperability is the key to smart-grid success. The envisioned smart-grid benefits can be realized only through appropriate levels of interoperability and the same can be ensured through adherence to appropriate standards. It is also important to have appropriate standards for communication systems and information technology tools to be interfaced effectively (Ayadi et al., 2019). There are so many institutes who have been working on development of standards for various aspects of smart-grid concept. Few such important developments in *standards for smart-grid and smart-grid interoperability as well as smart-grid communication and IT tools,* primarily those by National Institute of Standards and Technology (NIST) and National Renewable Energy Laboratory (NREL), have been listed out in Table 2.4 (NIST Framework and Roadmap for Smart Grid Interoperability Standards, Release 3.0, NIST Special Publication, n.d.)

Table 2.4 Standards for interoperability and communication and information exchange in smart grid.

Sn.	Standard	Application	SG conceptual architecture domain
1	ANSI C12.20	Measured data transfer through telephone networks	Customer, service providers
2	ANSI C12.21/IEEE P1702/MC1221	Optical interface of measuring units	Customer, service providers
3	ANSI/CEA 709 and Consumer Electronics Association (CEA) 852.1 LON Protocol Suite	LAN protocol for applications that include electric meters, street lighting, home and building automation, etc.	Customer, service providers
4	IEC 60870-6-503 Telecontrol Application Service Element 2 (TASE.2)	Exchange of messages between control centers of different utilities	Transmission, distribution

Continued

Table 2.4 Standards for interoperability and communication and information exchange in smart grid—cont'd

Sn.	Standard	Application	SG conceptual architecture domain
5	IEC 60870-6-702 Telecontrol Equipment and Systems	Defines a standard profile or set of options for implementing the application, presentation, and session layers	Transmission
6	IEC 60870-6-802 Telecontrol Equipment and Systems	Communications between electric power control centers	Transmission
7	IEC 61850 Suite	Communications within T&D substations for automation and protection	Transmission, distribution
8	IEC 61968/61970 Suites	Information exchange among control center systems using common information models. Also, define application-level energy management system interfaces and messaging for distribution grid management in the utility space	Operations
9	IEEE 1815 (DNP3) IEEE Xplore—IEEE Std 1815-2012	For automation of devices used in substations (SSs) and feeders. For intercommunication between control centers and SSs	Generation, transmission, distribution, operations, service provider
10	The IEEE Std 2030-2011	Smart Grid Interoperability	Transmission, distribution, operations
11	IEEE C37.118.2	Communications for phasor measurement units (PMUs)	Transmission, distribution
12	IEEE C37.238-2011 IEEE Standard Profile for Use of IEEE 1588 Precision Time Protocol in Power System Applications	Ethernet communications for power systems	Transmission, distribution

Table 2.4 Standards for interoperability and communication and information exchange in smart grid—cont'd

Sn.	Standard	Application	SG conceptual architecture domain
13	IEEE C37.239-2010	Interchange of power system event data	Transmission, distribution
14	IEEE 1547 Suite	Defines interconnections between the grid and distributed generation (DG) and storage	Transmission, distribution, customer
15	Inter-System Protocol (ISP)-based Broadband-Power Line Carrier (PLC) coexistence mechanism: (Portion of) IEEE 1901-2010 (ISP) and International Telecommunications Union Telecommunication Standardization Sector (ITU-T) G.9972 (06/2010) IEEE 1901-2010 ds/standard/1901-2010.html ITU-TG.9972	Both IEEE 1901-2010, "IEEE Standard for Broadband over Power Line Networks: Medium Access Control and Physical Layer Specifications," and ITU-TG.9972 (06/2010), "Coexistence mechanism for wireline home networking transceivers," specify Inter System Protocol (ISP)-based Broadband (>1.8 MHz) PLC (BB-PLC) coexistence mechanisms to enable the coexistence of different BBPLC protocols for home networking	Customer
16	NAESB REQ18, WEQ19 Energy Usage Information	Specifies two-way flows of energy usage information based on a standardized information model	Customer, service provider
17	OPC-UA Industrial	A platform-independent specification for a secure, reliable, high-speed data exchange based on a publish/subscribe mechanism	Customer
18	Open Automated Demand 2.0 Response (OpenADR)	The specification defines messages exchanged between the demand response (DR) service providers	Operations, service providers

Continued

Table 2.4 Standards for interoperability and communication and information exchange in smart grid—cont'd

Sn.	Standard	Application	SG conceptual architecture domain
19	Open Geospatial Consortium Geography Markup Language (GML)	A standard for exchange of location-based information addressing geographic data requirements for many smart-grid applications	Transmission, distribution
20	Organization for the Advancement of Structured Information Standard (OASIS) Energy Interoperation (EI)	Energy interoperation describes an information model and a communication model to enable demand response and energy transactions	Markets
21	Organization for the Advancement of Structured Information Standard (OASIS) EMIX (Energy Market Information eXchange)	EMIX provides an information model to enable the exchange of energy price, characteristics, time, and related information for wholesale energy markets, including market makers, market participants, quote streams, premises automation, and devices	Markets
22	Smart Energy Profile 2.0	Home Area Network (HAN) Device Communications and Information Model	Customer
23	Internet Protocol Suite, Request for Comments (RFC) 6272, Internet Protocols for the Smart Grid.	Internet Protocols for IP-based Smart Grid Networks IPv4/IPv6 are the foundation protocol for delivery of packets in the Internet network	Cross-cutting

(Basso & DeBlasio, 2012). Among these, IEEE 2030 and IEEE 1574 are the two noteworthy standards that dwell upon smart-grid interoperability and interconnection of various system components, respectively, and are listed in Table 2.5.

Financing the smart-grid deployment is a major challenge as going by the estimate as projected in Kuhn (2008), the cost of installing just the advanced metering infrastructure of the United States would range up to $27 billion and $1.5 trillion to upgrade the utility grid by 2030 (Chupka et al., 2008), notwithstanding the national regulations and policies that would allow the recovery of such huge investment. Smart grid supports integration of distributed renewable energy-sourced generation and hence characterized by bidirectional flow of energy and information. And AMIs further facilitate demand response activities providing better power quality and energy efficiency and consumer satisfaction. However, *developing economical storage*

Table 2.5 IEEE 1547 and IEEE 2030 standards for interoperability in a smart-grid.

The IEEE1547 series of standards (IEEE Std 1547 serves as the base standard of the IEEE 1547 series. Along with this series, seven complementary standards have also been designed to append the base standard. Five of these complementary standards have been published, and they are as listed below)	
IEEE Std 1547-2003 (reaffirmed 2008)	Standard for interconnecting DERs with EPSs
IEEE Std 1547.1-2005	Standard conformance test procedures for equipment interconnecting DERs with EPSs
IEEE Std 547.2-2008	Application guide for IEEE Std1547, IEEE standard for interconnecting DERs with EPSs
IEEE Std 1547.3-2007	Guide for monitoring, information exchange, and control of DERs interconnected with EPSs
IEEE Std 1547.4-2011	Guide for design, operation, and integration of DER island systems with EPSs
IEEE Std 1547.6-2011	Recommended practice for interconnecting DERs with EPSs distribution secondary network

IEEE standard 2030 series (The IEEE Std 2030-2011 Guide for smart grid interoperability; there are three additional complementary standards designed to expand upon the base 2030 standard and are as listed here)	
IEEE P2030.1	Guide for electric-sourced transportation infrastructure
IEEE P2030.2	Guide for the interoperability of energy storage systems integrated with the electric power infrastructure
IEEE P2030.3	Standard for test procedures for electric energy storage equipment and systems for electric power systems applications

system, which ensures above benefits, is crucial and another major technical challenge. Development and deployment of smart grid is a long-time affair, and during the course of this journey, *assimilating the changing technology, energy mixes, energy policy, and climate policies* could be a major impediment in smart grid deployment.

2.2 Issues and challenges with smart grid in the context of resiliency

Smart-grid inherently is conceptualized to be a resilient system, as the US Department of Energy in its report "Smart Grid System Report, July 2009" (U.S. Department of Energy, 2009) emphasizes "Operates resiliently to disturbances, attacks, and natural disasters" as one of the salient attributes of a smart-grid. However, during adoption of technologies such as micro-grid and AMIs and related paraphernalia for dynamic pricing and so on, the grid operation becomes too intricate, making it prone to instabilities and failures. On top of this, the impact of extreme climatic conditions enhances the likelihood of service disruptions and outages and a resilient system should have the capacity to recover from such disruption smoothly and efficiently. With this premise, resiliency of smart grid needs to be understood and analyzed in terms of "vulnerability" and "self-healing" (Das et al., 2020). Vulnerability of a system is due to its exposure to external environment arising scope for damage or loss. For example, the communication infrastructure layer exposes the physical infrastructure layer to the cyber space, making the grid vulnerable to cyberattack. The impact of various types of cyber-attack on grid operation is analyzed in An and Yang (2017), Shoukry et al. (2017), and Teixeira et al. (2015). This apart, there is plethora of LFHI events, both man-made and natural, which can play havoc on the physical infrastructure of the power system, the vulnerability in this aspect getting increased manifold due to heavy interdependence of the components in a smart grid. Based on the qualitative and qualitative assessment of the impact of these disastrous events, the smart grid should be designed to be adaptive to these situations, so as to be self-healing, a most required attribute of a resilient smart grid, which increases the post disaster recovery efficiency of the system. Self-healing property of a grid can be described as the ability of a grid whereby certain functionalities (generally those specific to critical infrastructures) get automatically restored by correction or reconfiguration of the system components. For example, self-healing by multiagent system based control (H. Liu, Chen, et al., 2012; S. Liu, Podmore, & Hou, 2012; Zidan & El-Saadany, 2012). However, to be self-healing, the system design must incorporate certain degree of "flexibility." Flexibility of a grid is a property that allows it to better respond to adverse events facilitating aforementioned auto-correction or reconfiguration of system components. For example, appliance-specific scheduling as discussed in Joe-Wong et al. (2012) is such a measure. It is interesting to note that microgrid technology, which is considered as a building block of smart grid, provides both flexibility and adaptability (i.e., self-healing) through intentional islanding. Thus, attributes such as vulnerability, self-healing, and recoverability, in addition to several others such as availability and logistics of components to be replaced, essentially constitute the resilience of the smart grid. And hence, study of resilience of smart grid requires a systematic analysis of several such components and their interactions under adverse scenarios.

Smart-grid being a vast and thickly interconnected dynamically evolving system of systems includes certain degree of operational randomness leading to instability and failures. Although tackling the operational intricacies is a challenge in itself, preparing the system to withstand and recover from incidences of vastly different and multiple numbers of adversarial events is the primary challenge in the development of a resilient smart grid.

Identifying framework and developing matrices for quantifying the response and recovery of such geographically distributed, thickly interconnected intricate system of systems is a major challenging issue. Table 2.6 provides a compilation of efforts made by various researchers in this aspect so far (G. Huang, Wang, Chen, Qi, & Guo, 2017).

Table 2.6 Efforts in the direction of formulation of framework and matrices for smart-grid resiliency (G. Huang, Wang, Chen, Qi, & Guo, 2017).

Authors and Ref.	System resiliency targeted at	Proposed framework/approach/ matrices to measure resiliency
Bruneau et al. (2003)	Earthquake	A generalized method to measure resilience against earthquake based on system's performance characteristics, quantified by measuring the extent of the anticipated degradation in quality over the time of recovery

Continued

Table 2.6 Efforts in the direction of formulation of framework and matrices for smart-grid resiliency (G. Huang, Wang, Chen, Qi, & Guo, 2017)—cont'd

Authors and Ref.	System resiliency targeted at	Proposed framework/approach/ matrices to measure resiliency
Tierney and Bruneau (2007)	Generalized	Based on a concept of the "*resilience triangle,*" the approach suggests two metrics of system resilience quantification: First one being the "infrastructure performance," which signifies the extent of loss to system functionality due to the occurred damage; the second is the "recovery time," which measures the manner in which the system gets restored over time
Cimellaro et al. (2010)	Generalized	The concept of "resilience triangle" was further expanded by suggesting that the hypotenuse of the resilience triangle could vary from a linear function and can take on forms such as exponential or trigonometric function
M. Panteli et al. (2017)	Generalized	"Resilience trapezoid (RT)" as an extension of "resilience triangle" was proposed to provide a comprehensive view of the level of critical infrastructure resilience during different phases of an event. The RT was formulated using the framework of metric, called "FLEP," comprising five different metrics, such as: (i) to what low (F) and (ii) at how rate (L) the resilience of the system decreases on the event of a disaster, (iii) how long (E) it stays in unrecovered state after the occurrence of the disaster, (iv) how fast (P) it bounces back to

Table 2.6 Efforts in the direction of formulation of framework and matrices for smart-grid resiliency (G. Huang, Wang, Chen, Qi, & Guo, 2017)—cont'd

Authors and Ref.	System resiliency targeted at	Proposed framework/approach/ matrices to measure resiliency
		its predisaster state, and finally (v) an area metric computed as the integral of the trapezoid to make an overall impact assessment
Petit et al. (2013)	Generalized	A resilience measurement index (RMI), developed to quantify the resilience critical infrastructures, uses a scale of 0 (low) to 100 (high) to provide resiliency comparisons. The RMI thus facilitates decision making in prioritization for improving resilience
2014 Quadrennial Energy Review (U.S. Department of Energy, 2014)	Generalized	The US Department of Energy (DOE) had conducted a workshop on "Resilience Metrics for Energy Transmission and Distribution Infrastructure" for the 2014 Quadrennial Energy Review (US Department of Energy, 2014) to identify and explore issues important to resilience quantification
Ji et al. (2017)	Generalized	Extensive discussion on fundamental challenges and advanced approaches on resilience quantification has been provided
Willis and Loa (2015)	Generalized	A comprehensive review on resilience metrics for energy systems has been presented.
G. Huang, Wang, Chen, Qi, and Guo (2017) and Bie et al. (2017)	Generalized	G. Huang, Wang, Chen, Qi, and Guo (2017) had quantified grid resilience as a measure of total load shed as against total load served. The larger the value of the denominator quantity, higher will be resilience of the system under study, as further enumerated by Bie et al. (2017)

The presence of communication or so-called cyber-layer interfaced to the physical layer makes the physical grid vulnerable to targeted attacks that can be carried out stealthily from remote locations. These threats lay active yet undetected for a long time before a major devastation could be inflicted through security breach. Cyber threat is a major issue in smart-grid resiliency perspective as it has wide repercussions ranging from system disruption to privacy breach of end consumers. Developing cyber security tools against ever smartening cyber hackers is a challenging task. Top ten possible smart-grid privacy concerns as enlisted by Dr. Christopher Velstos, proponent of Privacy Impact Assessment/NIST (IT Compliance), is provided in Table 2.7.

A disaster injured recovering grid could be further impacted due to cyberattacks, grossly hampering its resilience, which is one more and major challenges of multimodal adversities, with the capacity to provide a big blow to the recoverability of the grid.

Unpredictability in the severity and spatial coverage of an extremely adversarial event would all probability put all types of contingency measure haywire. Hence, deploying a resilient improvement plan and deciding upon the investment plan is another challenging issue.

The perceptions on resiliency do vary from local to regional to state. Hence, developing a holistic resilient measure addressing the geopolitical necessities is another challenge in smart-grid resilient enhancement.

Table 2.7 Smart-grid privacy concerns.

	Smart-grid privacy concerns
1	Theft identification
2	Figuring out pattern in an individual's activities
3	Getting to know the specific appliances used by the consumers/other stake holders
4	Carrying out real-time vigilance
5	Retrieving information on crucial activities through residual data
6	Possibility of invasions on individual residences
7	Possibility of providing information accidentally
8	Activity censorship
9	Possibility of making wrong decisions and subsequent actions taken based upon compromised data
10	Revealing activities when used with data from other utilities

3. Smart-grid resiliency enhancement

3.1 Smart grid resilience matrices

Before setting into any plan of deploying infrastructural augmentation for resiliency enhancement through efficient design and smooth operation of the smart grid, a systematic study based on a holistic framework is highly essential to develop tools for qualitative and quantitative assessment of system resilience against different disruptive events. Appropriate matrices thus developed provide the stakeholders, including the investors, means to decide upon the degree of resiliency needs to be adopted within the boundary of geopolitical and economic constraints.

3.1.1 Qualitative matrices

Literature is abound with several qualitative analyses that present frameworks and guidelines for analyzing and enhancing the resilience of the smart grid.

The authors in Molyneaux et al. (2016) dwelt upon ecology, psychology, risk management, and energy security as the factors of resilience assessment and proposed an interdisciplinary approach wherein adaptability, diversity, redundancy, system structure, and monitoring as common elements to be measured across all disciplines for resilience enhancement of a given system. The authors in Arghandeh et al. (2016) have deliberated on the challenges with a power system in the context of cyber-physical resilience. A resilient power system should be able to reconfigure its structure, loads, and resources in a seamless manner to maintain continuity of service under a given load prioritization. A. Sharifi and Yamagata (2015, 2016) have enlisted the characteristics of a resilient power system and proposed a conceptual framework for assessing the resilience of urban energy systems. Specifically, the authors identified *availability*, *accessibility*, *affordability*, and *acceptability* as four dimensions of resilience. Bian and Bie (2018) have proposed optimization model based on the coordination of multiple microgrids and have analyzed the performance of the system in terms of the extent of reduction in load. Carlson et al. (2012) have also proposed a conceptual framework for resiliency measurement identifying qualitative methods of measurement for critical infrastructure as well as community resilience. Similarly, another conceptual framework for power system resilience as proposed in M. Panteli and Mancarella (2015a, 2015b, 2017) is based on study of vulnerability and adaptability, identification and application of resilience improvement strategies, and cost-benefit analysis. The authors also proposed a short-term metrics which can be suitably modified as a long-term quantifying metrics for grid resilience. Along with strategies for hardening/reinforcing of the grids, improvement in operability measures and mixed strategies, the authors also presented an overall definition of critical infrastructures and used their conceptual framework and modeling tools to study the resilience of power systems (M. Panteli & Mancarella, 2015a, 2015b, 2017). Another conceptual framework that incorporates the temporal and spatial aspects as well as scales of the problem of quantifying resilience in the face of an energy crisis is presented in S. Erker et al. (2017a). The authors suggested merger of value-based and fact-based methods for assessing regional energy resilience to derive policy recommendations. The factual and value aspects of regional energy resilience were then applied to data collected in Austria (Exner et al., 2016) with respect to the temporal and spatial aspects of energy crisis (S. Erker et al., 2017b). A framework for organizing different metrics of resilience to guide both operational and strategic decisions, along with framework for categorization of metrics based on the scale of impact at facility/system level, system/region level, and region/nation level is provide by authors (Willis & Loa, 2015).

3.1.2 Quantitative metrics

One of the most popular and straight forward approach of quantifying the resilience of a system involves defining a quantity that is representative of the functional capabilities of the system. A quantity that varies across systems and is referred to in a common framework as the *Figure of Merit* (FOM) (Janić, 2018). FOM represents the functionality of a system in terms of the quantity (and quality) of the services delivered by the system. The evolution of FOM for a typical system in the presence of an extreme event is depicted in Fig. 2.4. Before the occurrence of a disastrous event, the system rests at a level of service denoted as FOM(−), which is less than the desired level of performance/service (FOM(*)). On occurrence of the disastrous event at time t1, the system performance experiences a sharp deterioration which settles down at the degraded value FOM(D) after time t2. The system undergoes recovery actions and settles at an improved performance level FOM(+). In Fig. 2.4, two different trajectories of system recovery are shown, which are results of different strategies adopted by the system, or different capabilities of the system. It can be observed from Fig. 2.4 that a system that follows the first trajectory

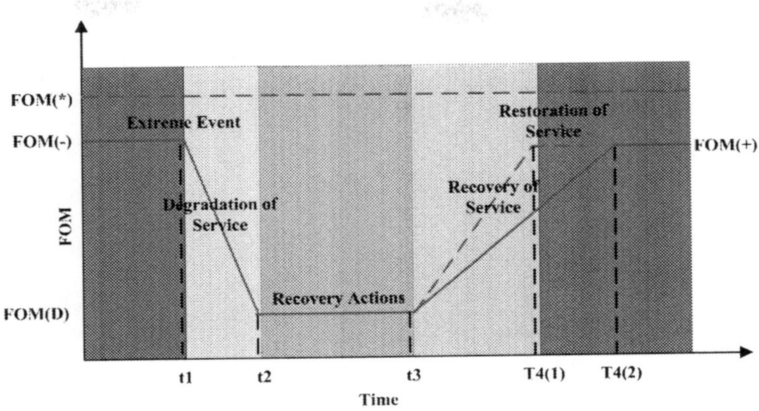

Fig. 2.4 Performance recovery (figure of merit) curve on a system on occurrence of an extreme adversarial event. *(No permission required.)*

recovers at time t4(1), whereas the one that follows the second trajectory takes longer and recovers at time t4(2), suggesting that the former is more resilient to the disastrous event than the latter. A detailed mathematical analysis on the formulation of dynamic resilience in this context can be referred from Das et al. (2020).

A summary of few major efforts in the direction of formulation of framework for smart-grid resiliency analysis and development of matrices is presented in Table 2.6.

3.2 Smart-grid resiliency enhancement

There have been numerous references to efforts made in designing and developing frameworks, guidelines, approaches, and modalities as well as tools for enhancement of smart-grid resilience. However, many of them are piece meal treatments, addressing issues of resilience against specific types of extreme events or focusing on specific geopolitical challenges or addressing resilience in specific service perspective. However, Huang et al. have provided a systematic study on all such attempts of resilience enhancement measures and have put them under a broad category of three important steps of smart-grid resiliency enhancement, namely resilience planning, resilience response, and resilience restoration (Fig. 2.5), which are based on the two major attributes of resiliency, i.e., adaptability and recoverability. Although resilience planning and resilience response are for enhancing system adaptability, resilience restoration step provides

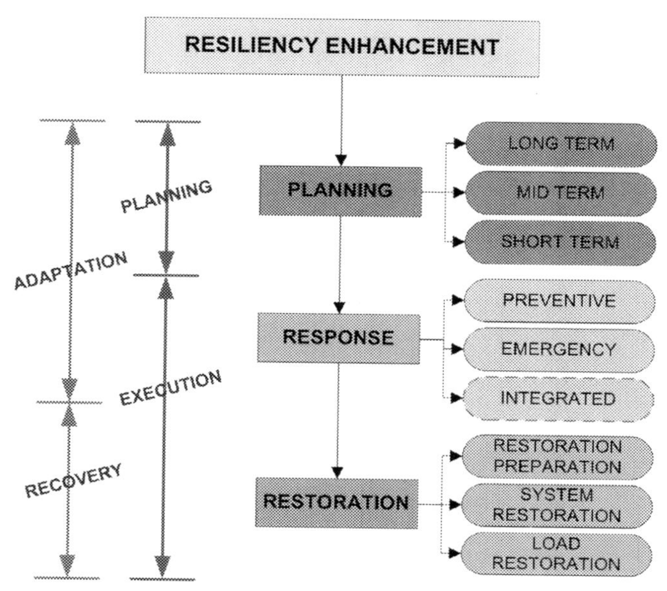

Fig. 2.5 Resilience enhancement strategy: Planning, execution, restoration. *(Modified from Huang, G., Wang, J., Chen, C., Qi, J., & Guo, C. (2017). Integration of preventive and emergency responses for power grid resilience enhancement,* IEEE Transactions on Power Systems, *99, 1.)*

the smart grid the ability to recover quickly post incidence of an extreme event. Although infrastructure hardening, vegetation management, and resource allocation are presented as the important steps under resilience planning stage, preventive response, emergency response, and integrated response are considered as three important resiliency enhancement measures under resilience response stage. Similarly, restoration preparation, system restoration, and load restoration are sequence of steps to be followed under resilience restoration stage. The study puts all major resilience enhancement proposals under appropriate segments in the afore described stages of resilience enhancement, and the same has been summarized in Table 2.8.

Table 2.8 A systematic smart-grid resilience enhancement approach (G. Huang, Wang, Chen, Qi, & Guo, 2017).

Resilience enhancement approaches	Steps to be undertaken	Methodologies adopted	Related research
Planning	**Infrastructure hardening** (The primary approach toward resilience enhancement planning)	Hardening strategies: • Upgradation of poles and towers by replacing aluminum structures with galvanized steel lattice or concrete in transmission systems, while replacing wooden poles with concrete, steel, or a suitable composite material in distribution systems (Executive Office of the President, 2013) • Alternative options include laying underground power lines (Yuan et al., 2016)	• A framework of hardening strategy based on pole failure and pole decay was proposed by Salman et al. (2015). Its effectiveness against hurricane was studied and analyzed • A tri-level optimization model as an effective hardening strategy in distribution system against extreme weather events was proposed in Ma et al. (2018)
	Vegetation management	• Trees and other flora and fauna below and around the rights of way of overhead transmission lines can cause major outages leading to cascading blackouts, and management of them by timely trimming these growths is known as vegetation management (Wanik et al.,	• A fixed-time interval schedule (traditional approach): trimming of vegetation on a scheduled interval based on pattern of growth of trees and weather studies (Dai & Christie, 1994) • Optimized scheduling for management of vegetation

Continued

Table 2.8 A systematic smart-grid resilience enhancement approach (G. Huang, Wang, Chen, Qi, & Guo, 2017)—cont'd

Resilience enhancement approaches	Steps to be undertaken	Methodologies adopted	Related research
		2017). It is also considered an important aspect of resilience planning • The North American Electric Reliability Corporation (NERC, 2012), in this context, prescribes the vegetation management standards for transmission owners to follow	around the distribution systems, worked out using a maintenance-scheduling algorithm (Kuntz et al., 2002)
	Resource allocation	Involves: • Optimized planning of power distribution • Allocations of DGs • Strategic placement of remotely controllable switches • Allocation and deputation of repair crew • Allocation of power electronics (PE) based controllers	• H. Gao et al. (2017): Prehurricane generating resources' allocation using a stochastic mixed-integer nonlinear program to have an optimized allocation plan for distribution systems • Yuan et al. (2016): Power line hardening through and allocation of DGs to minimize damage due to storm

	• Leveraging energy storage devices	• Xu et al. (2016): placement of remote controlled switches as a weighted set cover problem • Whipple (2014): A robust optimization model for prestorm repair crews' allocation with an objective of minimizing restoration time • Lei et al. (2016) A two-stage stochastic optimization model to allocate the truck-mounted mobile emergency generators before the occurrence of a natural disaster • Deployment of extra transmission lines (Wagaman, 2016) to have buffered back up of power • Allocation of PE-based controllers (Blaabjerg et al., 2017) for enhanced control on power flow • Leveraging energy storage devices (Wen et al., 2016) use of ESDs for supporting resilience response and restoration

Continued

Table 2.8 A systematic smart-grid resilience enhancement approach (G. Huang, Wang, Chen, Qi, & Guo, 2017)—cont'd

Resilience enhancement approaches	Steps to be undertaken	Methodologies adopted	Related research
Response	**Preventive responses** (actions available before disaster scenarios unfold	• Generator re-dispatch • Topology switching • Avoidance of load shedding	• Generator re-dispatch (G. Huang, Wang, Chen, Qi, & Guo, 2017; C. Wang et al., 2017) • Topology switching (G. Huang, Wang, Chen, Qi, & Guo, 2017), and adjustment of other facilities in the system • Notably, power grids in preventive state are operated in a way to meet all load demands without infringing any operational restrictions (Dy Liacco, 1967); thus, load shedding or similar actions cannot be resorted to
	Emergency responses (the actions taken immediately aftermath of a disaster)	System islanding/intentional islanding/controlled islanding is the most important emergency response strategy and is considered the last resort to prevent cascading blackouts	• Slow coherency analysis (You et al., 2004) and ordered binary decision diagrams (Sun et al., 2003) are traditional approaches of island identification

- For simultaneous multiple island formation, Fan et al. (2012) have proposed a mixed-integer programming approach to form islands in a power grid considering load shedding and connectivity constraints
- M. Golari et al. (2014) have formulated a two-stage stochastic programming model
- M. Golari et al. (2017) have further proposed a two-stage mixed-integer stochastic programming model to tackle extreme eventualities
- P. A. Trodden et al. (2013) have presented a mixed integer linear programming strategy to help decide upon deemed actions to be taken, in the context of load shedding/feeder isolation/generator switch off; a controlled islanding approach

Continued

Table 2.8 A systematic smart-grid resilience enhancement approach (G. Huang, Wang, Chen, Qi, & Guo, 2017)—cont'd

Resilience enhancement approaches	Steps to be undertaken	Methodologies adopted	Related research
			• An extended mixed-integer linear programming model was introduced by the same authors including voltage and reactive power constraints in their work (P. A. Trodden et al., 2014)
	Integrated response	Recent research works have identified that preventive and emergency responses can be used in tandem toward furtherance of system resilience (G. Huang, Wang, Chen, Qi, & Guo, 2017)	G. Huang, Wang, Chen, Qi, and Guo (2017) proposed an integrated resilience response (IRR) framework, which not only engages "situational awareness" for better resilience, but also provides effectual responses during both preventive and emergency stages. The IRR framework comprises two levels of robust mixed-integer optimization (RoMIO), which provides the optimal strategy for preventive response along with support in decision making during the emergency state

Restoration. Resilience restoration works toward quick restoration of power supply with minimized loss (Y. Wang, Chen, et al., 2016; Z. Wang, Wang, & Chen, 2016). The process generally involves stages; restoration preparation, system restoration, and load restoration (Adibi & Fink, 2006; Coffrin & Van Hentenryck, 2015; Fink et al., 1995)	**Restoration preparation**	Postdisaster restoration preparation includes steps such as: • Damage assessment, and taking stock of the availability of post-disaster resources • Restoration time estimation and fixing up the restoration strategy to be implemented (such as "bottom-up," "top-down," and hybrid strategy which is a combination of both) • Repair of critical infrastructures that are damaged and need to be made operated for system/load restoration • Off-line system restoration following the prescribed guidelines • System restoration using online tools, particularly in the event of unpredictable blackouts	There are a few repair scheduling problems identified in recent literatures; • A large-scale mixed nonlinear, nonconvex program based on two-stage approach by Coffrin and Van Hentenryck (2015) • A postdisaster repair scheduling problem for distribution system modeled on an integer linear programming using a multicommodity flow Tan et al. (2017) Some of the online strategies and guidelines: • "Development and Evaluation of System Restoration Strategies from a Blackout" (Final Report of Power Systems Engineering Research Center (PSERC) Project S30, 2009), which presents four modules; each performing specific tasks, such as (i) capacity optimization, (ii) transmission path search, (iii) constraint checking, and (iv) distribution system restoration
	System restoration works toward re-integrating the whole of the power system. Certain loads may also be restored during this stage with the intention of maintenance of system stability		

Continued

Table 2.8 A systematic smart-grid resilience enhancement approach (G. Huang, Wang, Chen, Qi, & Guo, 2017)—cont'd

Resilience enhancement approaches	Steps to be undertaken	Methodologies adopted	Related research
			• Generic Restoration Milestones (GRMs) as proposed by Hou et al. (2011) • A decision-support tool entitled "System Restoration Navigator" based on GRMs developed with the support of EPRI (S. Liu, Podmore, & Hou, 2012) • "Optimal Blackstart Capacity" (a tool also developed by EPRI) to ascertain the sufficiency of blackstart resources for system restoration (Jiang et al., 2017; Qiu et al., 2016)
	Load restoration (The last frontier of resilience enhancement)	• Once the system is restored back to sufficient strength, load restoration is the final battle to be won by the operators • Load restoration has become a critical and crucial stage of system restoration exercise mainly due to the increased level of DG penetration and embedded smart technologies in the present day smart grids	Research on load restoration in a smart grid context: • A mixed-integer linear programming model to maximize the critical loads to be picked up, using remote controlled switching devices and DGs Chen et al. (2016) • The microgrid-assisted critical load restoration problem as a two-objective chance constrained program under the uncertainties of renewable energy sources and load demand H. Gao et al. (2016)

- A mixed-integer nonlinear load restoration model to support the load restoration under reserve, frequency security, and steady-state constraints of power systems Qin et al. (2015)
- Effective use of DGs to overcome the cold load pickup (CLPU) problem Kumar et al. (2010)
- A three-phase microgrid restoration model to capture the unbalanced characteristics existing in distribution networks Y. Wang, Chen, et al. (2016) and Z. Wang, Wang, and Chen (2016)
- In addition, the majority of currently proposed approaches are worked out on radial topology, but scope lies in experimenting the same with meshed distribution networks (Heydt, 2010)

4. Conclusion

Resiliency of power systems is gaining increasing attention of researchers and corporate entities largely because of increasing number of outages due to climate change-induced disastrous events, which are also showing increasing trend of occurrence in recent times. At the same time, with radical change in demands and expectation of performance of power systems in terms of 21st-century parameters, coupled with consumers' participation through distributed generation and need for green energy, has brought about a paradigm shift in the way the new age power system network should evolve. What is now being termed as smart grid is essentially a gradually evolving grid through modernization. Though smart grid is envisaged and promises to be of immense benefits befitting of the modern lifestyle, there lies many challenges in its realization. Recent advancements in the field of communication, information technologies, and powerful computational technologies has indeed come as a boon in realization of smart grid. However, amalgamation of all these dynamic systems into the existing physical power system infrastructure not only makes the system complicated but also makes it hugely challenging in terms of operability, stability, and reliability. Deployment of smart-grid technologies in a live and functional existing power system is a fundamental challenge in itself. Though these issues have been largely addressed, the issue of resiliency in smart-grid context has a lot to look forward to. To begin with, to develop a resilient smart-grid system, a holistic framework is needed to be identified encompassing various geo-political-economic issues involving stakeholders across regional, national, and international levels. A generalized understanding and definition of resilience in smart-grid context is yet to be established. An effective and holistic approach of assessing the smart-grid resiliency and developing quantifying matrices is yet to reach a definitive level. Standards for design, development, and deployment of a resilient smart-grid addressing its various functionalities including interoperability of various intrasystems are tasks yet to reach its conclusive status. From a technical standpoint, several measures developed and proposed suffer from issues related with real-life implementation. Last but not the least, the cyber system, which is an integral subcomponent of a smart-grid makes it largely vulnerable to cyber threats, largely compromising its resiliency toward one of the important man-made disastrous event. Cybersecurity in a smart power infrastructure is still an area of research yet to be strengthened. So as, challenges are many so also are the opportunities.

References

Abbey, C. (2008). *Active distribution networks: Canadian example ProjectsCanadianExampleProject*. https://www.smartgrid.gov/files/documents/Active_Distribution_Networks_Canadian_Example_Projects_200812.pdf.

Adibi, M. M., & Fink, L. H. (2006). Overcoming restoration challenges associated with major power system disturbances—Restoration from cascading failures. *IEEE Power and Energy Magazine*, *4*(5), 68–77. https://doi.org/10.1109/MPAE.2006.1687819.

An, L., & Yang, G.-H. (2017). Secure state estimation against sparse sensor attacks with adaptive switching mechanism. *IEEE Transactions on Automatic Control*, *63*(8), 2596–2603.

Arghandeh, R., Von Meier, A., Mehrmanesh, L., & Mili, L. (2016). On the definition of cyber-physical resilience in power systems. *Renewable and Sustainable Energy Reviews*, *58*, 1060–1069. https://doi.org/10.1016/j.rser.2015.12.193.

Ayadi, F., Colak, I., & Bayindir, R. (2019). Interoperability in smart grid. In *7th International Conference on Smart Grid, icSmartGrid 2019* (pp. 165–169). Institute of Electrical and Electronics Engineers Inc. https://doi.org/10.1109/icSmartGrid48354.2019.8990680.

Basso, T., & DeBlasio, R. (2012). *IEEE smart grid series of standards IEEE 2030 (Interoperability) and IEEE 1547 (Interconnection) Status*.

Bian, Y., & Bie, Z. (2018). Multi-microgrids for enhancing power system resilience in response to the increasingly frequent natural hazards. *IFAC-PapersOnLine*, *51*(28), 61–66. https://doi.org/10.1016/j.ifacol.2018.11.678.

Bie, Z., Lin, Y., Li, G., & Li, F. (2017). Battling the extreme: A study on the power system resilience. *Proceedings of the IEEE*, *105*(7), 1253–1266. https://doi.org/10.1109/JPROC.2017.2679040.

Blaabjerg, F., Yang, Y., Yang, D., & Wang, X. (2017). Distributed power-generation systems and protection. *Proceedings of the IEEE*, *105*(7), 1311–1331. https://doi.org/10.1109/JPROC.2017.2696878.

Bruneau, M., Chang, S. E., Eguchi, R. T., Lee, G. C., O'Rourke, T. D., Reinhorn, A. M., Shinozuka, M., Tierney, K., Wallace, W. A., & Von Winterfeldt, D. (2003). A framework to quantitatively assess and enhance the seismic resilience of communities. *Earthquake Spectra*, *19*(4), 733–752. https://doi.org/10.1193/1.1623497.

Carlson, J., Haffenden, R., Bassett, G., Buehring, W., Collins, I. M., Folga, S., Petit, F., Phillips, J., Verner, D., & Whitfield, R. (2012). *Resilience: Theory and application*. Argonne National Lab (ANL).

Chandel, P., Thakur, T., & Sharma, R. (2015). Smart grid initiatives and experiences in India: Updates and review. *Vol. 7. Global Journal of Enterprise Information System* (4th, pp. 39–47). Global Journal of Enterprise Information System. http://www.gjeis.com/index.php/GJEIS/article/view/367/350. (Accessed 5 January 2022).

Chen, C., Wang, J., Qiu, F., & Zhao, D. (2016). Resilient distribution system by microgrids formation after natural disasters. *IEEE Transactions on Smart Grid*, 7(2), 958–966. https://doi.org/10.1109/TSG.2015.2429653.

Chupka, M., Earle, R., Fox-Penner, P., & Hledik, R. (2008). *Transforming America's power Industry: The investment challenge 2010-2030*. Prepared by The Brattle Group for The Edison Foundation.

Cimellaro, G. P., Reinhorn, A. M., & Bruneau, M. (2010). Framework for analytical quantification of disaster resilience. *Engineering Structures*, *32*(11), 3639–3649. https://doi.org/10.1016/j.engstruct.2010.08.008.

Coffrin, C., & Van Hentenryck, P. (2015). Transmission system restoration with co-optimization of repairs, load pickups, and generation dispatch. *International Journal of Electrical Power & Energy Systems*, *72*, 144–154. https://doi.org/10.1016/j.ijepes.2015.02.027.

Dai, L., & Christie, R. (1994). The optimal reliability-constrained line clearance scheduling problem. In *Vol. 1. Proceedings of the twenty-sixth annual North American power symposium* (pp. 244–249).
Das, L., Munikoti, S., Natarajan, B., & Srinivasan, B. (2020). Measuring smart grid resilience: Methods, challenges and opportunities. *Renewable and Sustainable Energy Reviews, 130*. https://doi.org/10.1016/j.rser.2020.109918.
Department of Energy, United States (DOE). (2003). *Grid 2030—A vision for electricity's second 100 years*. https://www.energy.gov/sites/default/files/oeprod/DocumentsandMedia/Electric_Vision_Document.pdf.
Department of Energy, United States (DOE). (2008). *The smart grid: An introduction*. https://www.energy.gov/sites/default/files/oeprod/DocumentsandMedia/DOE_SG_Book_Single_Pages%281%29.pdf.
Dy Liacco, T. E. (1967). The adaptive reliability control system. *IEEE Transactions on Power Apparatus and Systems, 86*(5), 517–531. https://doi.org/10.1109/TPAS.1967.291728.
E.T.P. (2006). *European technology platform smart grids: Vision and strategy for Europe's electricity networks of the future*.
EPRI. (2010). *EPRI smart grid demonstration initiative: Final update*. https://smartgrid.epri.com/doc/EPRI%20Smart%20Grid%20Demonstration%202-Year%20Update_final.pdf.
EPRI. (2021). *Retrieved August 1, 2021, from*. http://smartgrid.epri.com/.
Erker, S., Stangl, R., & Stoeglehner, G. (2017a). Resilience in the light of energy crises—Part I: A framework to conceptualise regional energy resilience. *Journal of Cleaner Production, 164*, 420–433. https://doi.org/10.1016/j.jclepro.2017.06.163.
Erker, S., Stangl, R., & Stoeglehner, G. (2017b). Resilience in the light of energy crises—Part II: Application of the regional energy resilience assessment. *Journal of Cleaner Production, 164*, 495–507. https://doi.org/10.1016/j.jclepro.2017.06.162.
Executive Office of the President. (2013). *Economic benefits of increasing electric grid resilience to weather outages*. https://www.energy.gov/sites/default/files/2013/08/f2/Grid%20Resiliency%20Report_FINAL.pdf.
Exner, A., Politti, E., Schriefl, E., Erker, S., Stangl, R., Baud, S., Warmuth, H., Matzenberger, J., Kranzl, L., Paulesich, R., Windhaber, M., Supper, S., & Stoeglehner, G. (2016). Measuring regional resilience towards fossil fuel supply constraints. Adaptability and vulnerability in socio-ecological transformations-the case of Austria. *Energy Policy, 91*, 128–137. https://doi.org/10.1016/j.enpol.2015.12.031.
Fan, N., Izraelevitz, D., Pan, F., Pardalos, P. M., & Wang, J. (2012). A mixed integer programming approach for optimal power grid intentional islanding. *Energy Systems, 3*(1), 77–93. https://doi.org/10.1007/s12667-011-0046-5.
Farhangi, H. (2010). The path of the smart grid. *IEEE Power and Energy Magazine, 8*(1), 18–28. https://doi.org/10.1109/MPE.2009.934876.
Federation, G. S. G. (2012). *Global smart grid federation report*. Global Smart Grid Federation.
Fink, L. H., Liou, K. L., & Liu, C. C. (1995). From generic restoration actions to specific restoration strategies. *IEEE Transactions on Power Systems, 10*(2), 745–752. https://doi.org/10.1109/59.387912.
Gao, H., Chen, Y., Mei, S., Huang, S., & Xu, Y. (2017). Resilience-oriented pre-hurricane resource allocation in distribution systems considering electric buses. *Proceedings of the IEEE, 105*(7), 1214–1233. https://doi.org/10.1109/JPROC.2017.2666548.
Gao, H., Chen, Y., Xu, Y., & Liu, C. C. (2016). Resilience-oriented critical load restoration using microgrids in distribution systems. *IEEE Transactions on Smart Grid, 7*(6), 2837–2848. https://doi.org/10.1109/TSG.2016.2550625.
Gellings, C. W. (2009). *The smart grid: Enabling energy efficiency and demand response*.
Gilbert, E. I., Violette, D. M., & Rogers, B. (2011). *Paths to smart grid interoperability*.

Golari, M., Fan, N., & Wang, J. (2014). Two-stage stochastic optimal islanding operations under severe multiple contingencies in power grids. *Electric Power Systems Research, 114*, 68–77. https://doi.org/10.1016/j.epsr.2014.04.007.

Golari, M., Fan, N., & Wang, J. (2017). Large-scale stochastic power grid islanding operations by line switching and controlled load shedding. *Energy Systems, 8*(3), 601–621. https://doi.org/10.1007/s12667-016-0215-7.

Haravi, H., & Ghafurian, R. (2011). Smart grid: The electric energy system of the future. *99. Proceedings of the IEEE* (6th, pp. 917–921). IEEE. https://doi.org/10.1109/JPROC.2011.2124210.

Herter, K., O'Connor, T., & Navarro, L. (2011). *Evaluation framework for smart grid deployment plans a systematic approach for assessing plans to benefit customers and the environment*. Herter, K., O'Connor, T. and Navarro, L., 2010. Evaluation Framework for Smart Grid Deployment Plans. Herter Consulting-EDF. https://www.smartgrid.gov/files/documents/Evaluation_Framework_for_Smart_Grid_Deployment_Plan_201112.pdf. (Accessed 5 January 2022).

Heydt, G. T. (2010). The next generation of power distribution systems. *IEEE Transactions on Smart Grid, 1*(3), 225–235. https://doi.org/10.1109/TSG.2010.2080328.

Hou, Y., Liu, C. C., Sun, K., Zhang, P., Liu, S., & Mizumura, D. (2011). Computation of milestones for decision support during system restoration. *IEEE Transactions on Power Systems, 26*(3), 1399–1409. https://doi.org/10.1109/TPWRS.2010.2089540.

Huang, G., Wang, J., Chen, C., Qi, J., & Guo, C. (2017). Integration of preventive and emergency responses for power grid resilience enhancement. *IEEE Transactions on Power Systems, 99*, 1.

Industry 4.0. (2021). https://www.i-scoop.eu/industry-4-0/smart-grids-electrical-grid/.

Janić, M. (2018). Modelling the resilience of rail passenger transport networks affected by large-scale disruptive events: the case of HSR (high speed rail). *Transportation, 45*(4), 1101–1137. https://doi.org/10.1007/s11116-018-9875-6.

Ji, C., Wei, Y., & Poor, H. V. (2017). Resilience of energy infrastructure and services: Modeling, data analytics, and metrics. *Proceedings of the IEEE, 105*(7), 1354–1366. https://doi.org/10.1109/JPROC.2017.2698262.

Jiang, Y., Chen, S., Liu, C. C., Sun, W., Luo, X., Liu, S., Bhatt, N., Uppalapati, S., & Forcum, D. (2017). Blackstart capability planning for power system restoration. *International Journal of Electrical Power & Energy Systems, 86*, 127–137. https://doi.org/10.1016/j.ijepes.2016.10.008.

Joe-Wong, C., Sen, S., Ha, S., & Chiang, M. (2012). Optimized day-ahead pricing for smart grids with device-specific scheduling flexibility. *IEEE Journal on Selected Areas in Communications, 30*(6), 1075–1085. https://doi.org/10.1109/JSAC.2012.120706.

Jones, P. (2010). The role of new technologies: A power engineering equipment supply base perspective. In *Grid policy workshop*.

Kabalci, E., & Kabalci, Y. (2019). Introduction to smart grid architecture. In E. Kabalci, & Y. Kabalci (Eds.), *Smart grids and their communication systems* Springer. https://doi.org/10.1007/978-981-13-1768-2_1.

Kappagantu, R., & Daniel, S. A. (2018). Challenges and issues of smart grid implementation: A case of Indian scenario. *Journal of Electrical Systems and Information Technology*, 453–467. https://doi.org/10.1016/j.jesit.2018.01.002.

Kappagantu, R., Daniel, S. A., & Suresh, N. S. (2016). Techno-economic analysis of smart grid pilot project—Puducherry. *Resource-Efficient Technologies*, 185–198. https://doi.org/10.1016/j.reffit.2016.10.001.

Kuhn, T. R. (2008). Legislative proposals to reduce greenhouse gas emissions: An overview. *Testimony before the United States house of representatives subcommittee on energy and air quality*. United States house of representatives subcommittee. http://energycommerce.house.gov/Press_110/110st177.shtml. (Accessed 5 January 2022).

Kumar, V., Kumar, H. C. R., Gupta, I., & Gupta, H. O. (2010). DG integrated approach for service restoration under cold load pickup. *IEEE Transactions on Power Delivery, 25*(1), 398–406. https://doi.org/10.1109/TPWRD.2009.2033969.

Kuntz, P. A., Christie, R. D., & Venkata, S. S. (2002). Optimal vegetation maintenance scheduling of overhead electric power distribution systems. *IEEE Transactions on Power Delivery, 17*(4), 1164–1169. https://doi.org/10.1109/TPWRD.2002.804007.

Lei, S., Wang, J., Chen, C., & Hou, Y. (2016). Mobile emergency generator pre-positioning and real-time allocation for resilient response to natural disasters. *IEEE Transactions on Smart Grid*, 1. https://doi.org/10.1109/TSG.2016.2605692.

Liu, H., Chen, X., Yu, K., & Hou, Y. (2012). The control and analysis of self-healing urban power grid. *IEEE Transactions on Smart Grid, 3*(3), 1119–1129. https://doi.org/10.1109/TSG.2011.2167525.

Liu, S., Podmore, R., & Hou, Y. (2012). System restoration navigator: A decision support tool for system restoration. In *IEEE power and energy society general meeting*. https://doi.org/10.1109/PESGM.2012.6344826.

Ma, S., Chen, B., & Wang, Z. (2018). Resilience enhancement strategy for distribution systems under extreme weather events. *IEEE Transactions on Smart Grid*, 1442–1451. https://doi.org/10.1109/TSG.2016.2591885.

Miller, J. (2008). *The smart grid–How do we get there*. Smart Grid News.

Molyneaux, L., Brown, C., Wagner, L., & Foster, J. (2016). Measuring resilience in energy systems: Insights from a range of disciplines. *Renewable and Sustainable Energy Reviews, 59*, 1068–1079. https://doi.org/10.1016/j.rser.2016.01.063.

NERC. (2012). *FAC-003-3-transmission vegetation management*. https://www.nerc.com/files/FAC-003-3.pdf.

NIST framework and roadmap for smart grid interoperability standards, release 3.0, NIST special publication. (n.d.).

OECD. (2009). *Smart sensor networks: Technologies and applications f or green growth*. https://www.oecd.org/sti/44379113.pdf.

Panteli, M., & Mancarella, P. (2015a). Influence of extreme weather and climate change on the resilience of power systems: Impacts and possible mitigation strategies. *Electric Power Systems Research, 127*, 259–270. https://doi.org/10.1016/j.epsr.2015.06.012.

Panteli, M., & Mancarella, P. (2015b). The grid: Stronger, bigger, smarter?: Presenting a conceptual framework of power system resilience. *IEEE Power and Energy Magazine, 13*(3), 58–66. https://doi.org/10.1109/MPE.2015.2397334.

Panteli, M., & Mancarella, P. (2017). Modeling and evaluating the resilience of critical electrical power infrastructure to extreme weather events. *IEEE Systems Journal, 11*(3), 1733–1742. https://doi.org/10.1109/JSYST.2015.2389272.

Panteli, M., Mancarella, P., Trakas, D. N., Kyriakides, E., & Hatziargyriou, N. D. (2017). Metrics and quantification of operational and infrastructure resilience in power systems. *IEEE Transactions on Power Systems, 32*(6), 4732–4742. https://doi.org/10.1109/TPWRS.2017.2664141.

Patnaik, B., Mishra, M., Bansal, R. C., & Jena, R. K. (2020). AC microgrid protection–A review: Current and future prospective. *Applied Energy, 271*.

Petit, F., Bassett, G. W, Black, R., Buehring, W. A., Collins, M. J., Dickinson, D. C., … Idaho National Laboratory. (2013). Resilience measurement index: An indicator of critical infrastructure resilience. *Technical Report, ANL/DIS-13-01, DE-AC02-06CH11357*. United States: Argonne National Lab. (ANL), Argonne, IL. https://doi.org/10.2172/1087819. (Accessed 5 January 2022).

Qin, Z., Hou, Y., Liu, C. C., Liu, S., & Sun, W. (2015). Coordinating generation and load pickup during load restoration with discrete load increments and reserve constraints. *IET Generation, Transmission and Distribution, 9*(15), 2437–2446. https://doi.org/10.1049/iet-gtd.2015.0240.

Qiu, F., Wang, J., Chen, C., & Tong, J. (2016). Optimal black start resource allocation. *IEEE Transactions on Power Systems*, *31*(3), 2493–2494. https://doi.org/10.1109/TPWRS.2015.2442918.

Salman, A. M., Li, Y., & Stewart, M. G. (2015). Evaluating system reliability and targeted hardening strategies of power distribution systems subjected to hurricanes. *Reliability Engineering and System Safety*, *144*, 319–333. https://doi.org/10.1016/j.ress.2015.07.028.

Sharifi, A., & Yamagata, Y. (2015). A conceptual framework for assessment of urban energy resilience. In *Vol. 75. Energy procedia* (pp. 2904–2909). Elsevier Ltd. https://doi.org/10.1016/j.egypro.2015.07.586.

Sharifi, A., & Yamagata, Y. (2016). Principles and criteria for assessing urban energy resilience: A literature review. *Renewable and Sustainable Energy Reviews*, *60*, 1654–1677. https://doi.org/10.1016/j.rser.2016.03.028.

Shoukry, Y., Nuzzo, P., Puggelli, A., Sangiovanni-Vincentelli, A. L., Seshia, S. A., & Tabuada, P. (2017). Secure state estimation for cyber-physical systems under sensor attacks: A satisfiability modulo theory approach. *IEEE Transactions on Automatic Control*, *62*(10), 4917–4932. https://doi.org/10.1109/TAC.2017.2676679.

Smart Energy Consumer Collaborative. (2013). *Smart grid economic and environmental benefits*. https://smartenergycc.org/smart-grid-economic-and-environmental-benefits-report/.

Sun, K., Zheng, D. Z., & Lu, Q. (2003). Splitting strategies for islanding operation of large-scale power systems using OBDD-based methods. *IEEE Transactions on Power Systems*, *18*(2), 912–923. https://doi.org/10.1109/TPWRS.2003.810995.

Tan, Y., Qiu, F., Das, A. K., Kirschen, D. S., Arabshahi, P., & Wang, J. (2017). Scheduling post-disaster repairs in electricity distribution networks. *arXiv*. https://arxiv.org.

Teixeira, A., Sou, K. C., Sandberg, H., & Johansson, K. H. (2015). Secure control systems: A quantitative risk management approach. *IEEE Control Systems*, *35*(1), 24–45. https://doi.org/10.1109/MCS.2014.2364709.

Tierney, K., & Bruneau, M. (2007). *Conceptualizing and measuring resilience: A key to disaster loss reduction*. TR News. https://onlinepubs.trb.org/onlinepubs/trnews/trnews250_p14-17.pdf. (Accessed 5 January 2022).

Ton, D. (2009). DOE's perspectives on smart grid technology, challenges, & research opportunities. In *UCLA engineering smartgrid seminar*.

Trodden, P. A., Bukhsh, W. A., Grothey, A., & McKinnon, K. I. M. (2013). MILP formulation for controlled islanding of power networks. *International Journal of Electrical Power & Energy Systems*, *45*(1), 501–508. https://doi.org/10.1016/j.ijepes.2012.09.018.

Trodden, P. A., Bukhsh, W. A., Grothey, A., & McKinnon, K. I. M. (2014). Optimization-based Islanding of power networks using piecewise linear AC power flow. *IEEE Transactions on Power Systems*, *29*(3), 1212–1220. https://doi.org/10.1109/TPWRS.2013.2291660.

U.S. Department of Energy. (2009). *Smart grid system report*. https://www.energy.gov/sites/default/files/2009%20Smart%20Grid%20System%20Report.pdf.

U.S. Department of Energy. (2014). *Technical workshop: Resilience metrics for energy transmission and distribution infrastructure*. https://www.energy.gov/sites/prod/files/2015/01/f19/QER%20Workshop%20June%2010%202014%20Posted.pdf.

US Department of Electricity. (2012). *Demand reduction from application of advance metering infrastructure, pricing programs, and customer-based systems—Initial results*. https://www.energy.gov/sites/prod/files/2016/10/f33/peak_demand_report_final_12-13-2012_0.pdf.

US Department of Electricity. (2013). *Smart grid 2013 global impact report*. https://www.smartgrid.gov/files/documents/global_smart_grid_impact_report_2013.pdf.

Wagaman, A. (2016). *PPL nearing completion of backup transmission line in Emmaus*. https://www.mcall.com/news/local/east-penn/mc-emmaus-new-ppl-transmission-line-20160119-story.html.

Wang, J., Biviji, M., & Wang, W. M. (2011). Case studies of smart grid demand response programs in North America. In *IEEE PES innovative smart grid technologies conference Europe, ISGT Europe*. https://doi.org/10.1109/ISGT.2011.5759162.

Wang, Y., Chen, C., Wang, J., & Baldick, R. (2016). Research on resilience of power systems under natural disasters—A review. *IEEE Transactions on Power Systems*, *31*(2), 1604–1613. https://doi.org/10.1109/TPWRS.2015.2429656.

Wang, C., Hou, Y., Qiu, F., Lei, S., & Liu, K. (2017). Resilience enhancement with sequentially proactive operation strategies. *IEEE Transactions on Power Systems*, *32*(4), 2847–2857. https://doi.org/10.1109/TPWRS.2016.2622858.

Wang, Z., Wang, J., & Chen, C. (2016). A three-phase microgrid restoration model considering unbalanced operation of distributed generation. *IEEE Transactions on Smart Grid*, *99*(1), 1–11.

Wanik, D. W., Parent, J. R., Anagnostou, E. N., & Hartman, B. M. (2017). Using vegetation management and LiDAR-derived tree height data to improve outage predictions for electric utilities. *Electric Power Systems Research*, *146*, 236–245. https://doi.org/10.1016/j.epsr.2017.01.039.

Wen, Y., Li, W., Huang, G., & Liu, X. (2016). Frequency dynamics constrained unit commitment with battery energy storage. *IEEE Transactions on Power Systems*, *31*(6), 5115–5125. https://doi.org/10.1109/TPWRS.2016.2521882.

Whipple, S. D. (2014). *Predictive storm damage modeling and optimizing crew response to improve storm response operations*. Thesis S.M., Massachusetts Institute of Technology, Engineering Systems Division. https://dspace.mit.edu/handle/1721.1/90166.

Willis, H. H., & Loa, K. (2015). *Measuring the resilience of energy distribution systems*.

Xu, Y., Liu, C. C., Schneider, K. P., & Ton, D. T. (2016). Placement of remote-controlled switches to enhance distribution system restoration capability. *IEEE Transactions on Power Systems*, *31*(2), 1139–1150. https://doi.org/10.1109/TPWRS.2015.2419616.

You, H., Vittal, V., & Wang, X. (2004). Slow coherency-based islanding. *IEEE Transactions on Power Systems*, *19*(1), 483–491. https://doi.org/10.1109/TPWRS.2003.818729.

Yuan, W., Wang, J., Qiu, F., Chen, C., Kang, C., & Zeng, B. (2016). Robust optimization-based resilient distribution network planning against natural disasters. *IEEE Transactions on Smart Grid*, 7(6), 2817–2826. https://doi.org/10.1109/TSG.2015.2513048.

Zidan, A., & El-Saadany, E. F. (2012). A cooperative multiagent framework for self-healing mechanisms in distribution systems. *IEEE Transactions on Smart Grid*, *3*(3), 1525–1539. https://doi.org/10.1109/TSG.2012.2198247.

CHAPTER THREE

Remedial action scheme to improve resiliency under failures in the Central American power grid

Wilfredo C. Flores[a], Javier Barrionuevo[b], and Santiago P. Torres[c]

[a]Engineering Faculty, Universidad Tecnológica Centroamericana (UNITEC), Tegucigalpa, Honduras
[b]Center for Studies of Energy Regulatory Activity, University of Buenos Aires (UBA), Argentina
[c]Department of Electrical, Electronics and Telecommunications Engineering, University of Cuenca (UC), Cuenca, Ecuador

1. Introduction

Remedial action schemes (RAS), also known as special protection systems (SPS), have been widely used to provide protection for power systems against problems not directly involving specific equipment fault protection (WECC, 1996).

RAS are designed to detect abnormal system conditions and take predetermined, corrective action to preserve system integrity and provide acceptable system performance. Today, in many parts of the world, RAS represent an economic, smart, and viable planning and operational alternative to extending transmission system capability (McCalley, 2010). Thus, it has been demonstrated in practice that RAS make the operation of any electrical network more robust, reliable, and smarter. In this sense and within the framework of the smart grids, it is well known that this kind of technologies are helping utilities to speed outage restoration following major extreme events, reduce the total number of affected customers, and improve overall service reliability to reduce customer losses from power disruptions (U.S. Department of Energy, 2014).

In Central America: Costa Rica, El Salvador, Guatemala, Honduras, Nicaragua, and Panama, as part of the Central American Integration System, have developed a competitive electrical regional market by interconnecting their national networks through transmission links and promoting regional generation projects to develop the electricity industry for the benefit of the region's inhabitants. In the same way, the Central American power

Electric Power Systems Resiliency
https://doi.org/10.1016/B978-0-323-85536-5.00009-6

transmission grid is interconnected with the Mexican electrical power grid (W.C. Flores, 2012).

In this sense, energy exchanges among countries have been increasing since the creation of the Regional Electric Power Market, which has caused greater power flows in some power lines of the current regional electric power system. However, the regional transmission grid still has constraints that are shown in Section 2 of this document, so the implementation of an RAS is a viable option that allows intelligent improvement of the operation and resiliency increase of the electric grid without investing vast resources. The latter is broadly showed by W.C. Flores et al. (2017).

Hence, this work shows an RAS implementation whose technical viability is shown through simulation using four failure scenarios located in several points of the Central American electric power grid. It is demonstrated that there can be an improvement in the performance of an electric power grid through the implementation of an RAS when the electric grid is composed by electric power systems from different countries, with different electrical realities, which is known as a Regional Electric Power Grid. Hence, the major contributions of this work are the following:

This work demonstrates that power grid blackouts can be avoided using regional protections and that reliable operation of different control areas is possible.

This chapter is organized as follows: in Section 2, the problems of the Regional Power grid are shown; Section 3 shows the proposed RAS's operating premises and flow chart; in Section 4, the simulation of the RAS under four different faults are shown; finally, in Section 5, the conclusions are presented.

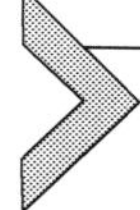

2. Central American power grid

2.1 Radial configuration

Because of its geographic nature, the Central American power grid has a radial topology (Fig. 3.1). It is well known that this type of radial configuration has low reliability compared to a mesh network (Gómez-Expósito et al., 2009). Thus, without proper protection, a failure on certain important links could lead the network to critical operating conditions. Also, under fault conditions, the opening of certain lines could cause islands, so that the reliability of the network could become critical under extreme operating conditions and without adequate protection and monitoring measures.

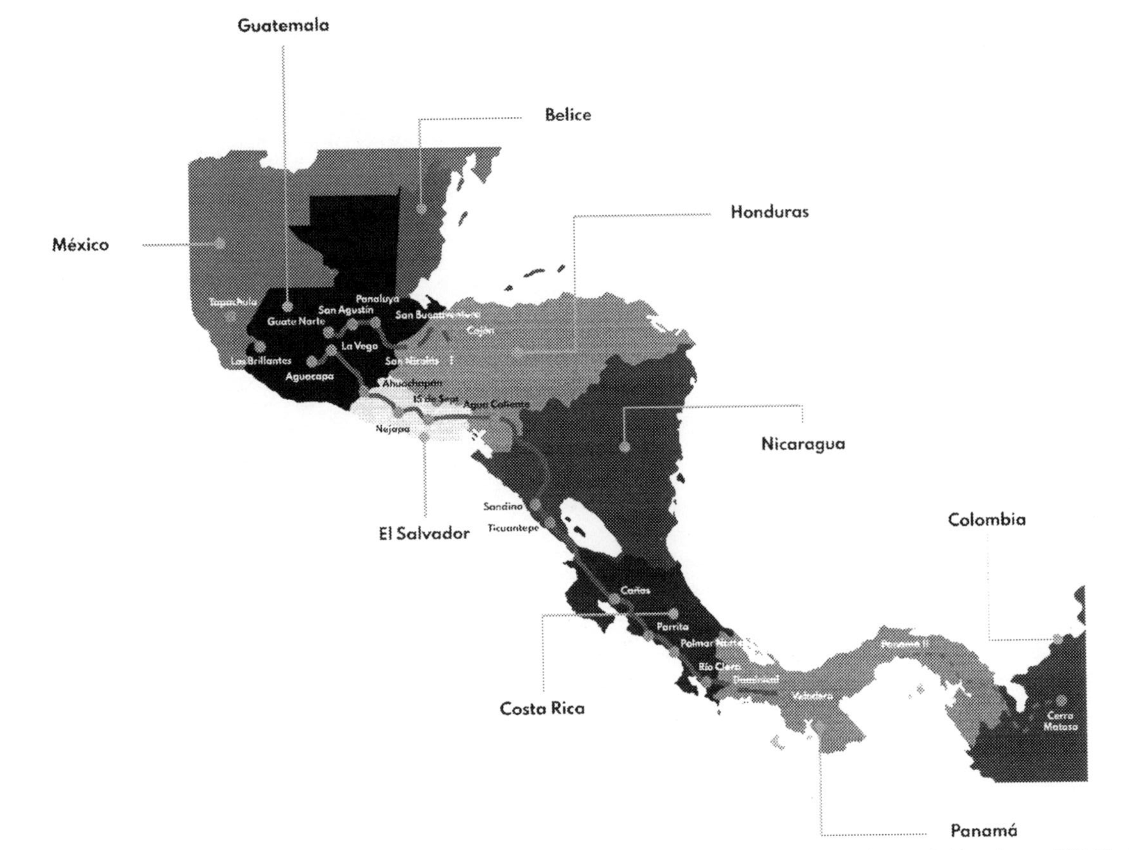

Fig. 3.1 Central American electric power grid (230kV). *CRI*, Costa Rica; *ES*, El Salvador; *GUA*, Guatemala; *HND*, Honduras; *MX*, Mexico; *NI*, Nicaragua; *PA*, Panamá. *(No permission required.)*

2.2 Constraints in the capacities of lines and power transformers

As in any electrical power grid, devices, both lines and transformers, have technical limitations which depend on their operative values obtained from security analysis. In the case of the Central American power grid, there are links that could lead to critical operating conditions in the event of a failure. Although there are national electricity networks considered to be reliably connected to the Central American Regional Transmission Power Grid, there are other National networks whose reliability is considered low, so a failure in the latter could lead the rest of the countries to critical conditions. In the same way, it is necessary to take into account the nominal capacities of certain important power transformers in the network. These power transformation capacities are limited in countries such as Honduras, where a large quantity of power transformers in their national network have been congested for a long time (AETS, BCEOM and EDE Ingenieros, 2010), which also limits transfer capacities.

2.3 Reactive power compensation

Reactive power compensation is necessary to improve transfers between countries. It is well known that in an electrical network the reactive power is used to avoid voltage problems along the transmission network. Without adequate reactive compensation, transfers between countries are smaller according to the transfer capacity of lines.

Although the tie to Mexico has helped in stabilizing the system under critical conditions, it has been observed that the increased power flow coming from Mexico, while crossing a country to supply the load unbalance of another country, has caused an increased consumption of reactive power in the country that is being crossed over. This causes a dramatic voltage drop in the transfer path of the country which forces to trip the tie-lines, to avoid its own voltage collapse. These circumstances can be prevented by applying intelligent load shedding schemes based on either RAS or phasor measurement units (PMU) (Sanchez et al., 2011).

2.4 Interarea oscillations

Additionally, intersystem oscillations have been frequently observed in the Mexican power system (E. Martinez & Messina, 2008; E.M. Martinez et al., 2013), with which the problem of the interarea oscillations becomes critical for the Central American power system. Interarea oscillations are associated with machines in one part of the system oscillating against machines in other part of the system. They are caused for two or more groups of closely coupled machines being interconnected by weak ties. The natural frequency of these oscillations is typically in the range of 0.1–1 Hz (Grigsby, 2001).

The small signal stability problem of a power system occurs usually as a result of insufficient damping of electromechanical oscillations (Kundur et al., 2004). The topology of the Central American power grid is prone to inadequate damping or negative damping of small signal, which generates interarea oscillations. In the past, these kinds of oscillations have occurred between hydroelectric power plants located in Central America, forcing the installation of power system stabilizers (PSS) in the participating generators (Sanchez et al., 2011). After the interconnection with Mexico, new interarea oscillations have emerged in the system (W.C. Flores, 2016). For handling such cases, PMU can provide remote measurements which can be used for re-tuning the PSSs and to improve their effectiveness. Also, wide area measurement systems (WAMS) can also be used for developing a coordinated type of control using PSS, static var. compensators (SVC), etc. to damp all low-frequency oscillations, as shown by Vance et al. (2012). To date, important quantities of PMU are being installed in the Central American power grid (W. Flores et al., 2012). In the future, the small signal stability problem in Central America power grid could increase because of the rise of wind and PV generation in the region (Rueda & Shewarega, 2009).

Following, the operating premises and flowchart of the RAS proposed in this work are shown, which helps to avoid some of the problems discussed above. It is worth mentioning that the RAS proposed in this work was designed to resolve a particular critical condition, that is, for an Ad-hoc function.

3. Operating premises and flowchart of the RAS

Before showing the RAS's flowchart proposed in this work, it is necessary to show the premises with which it operates. In this sense, it is important to mention that the goal of this RAS is to keep the power grid stable under large outage of power generation caused by failures, with which the resilience of the grid is raised.

3.1 Operating premises for the RAS

The RAS will operate under the following premises:

- The RAS operation occurs when the conditions are met according to the coordination of the protection adapted to the hierarchical protection systems (operation coordinated with other protection schemes already installed in the network).
- If during the operation of the RAS islands are formed in the grid, these must be stable.
- During the event of power generation outage due to fault in the systems south of Guatemala, to avoid overloading of the power transformer that connects Mexico and Guatemala, due to the flow by the response of the Mexican power system.
- Under the event of power generation outage as a result of fault in the systems south of Nicaragua, to avoid overloading of main lines in Nicaragua, which could affect the rest of the Regional Power System.
- For a time of $t = t1$, the power flow in the Guatemala-El Salvador link should not exceed a maximum value of *Pmx1*. If it exceeds this value, then the Guatemala-El Salvador link opens.
- For a time $t = t2$, the power flow in the Honduras-Nicaragua link should not exceed a maximum value of *Pmx2*. Also, at the same time, the power flow through Nicaragua's main lines should not exceed a maximum value of *Pmx3*. If it exceeds this value, then the Honduras-Nicaragua link opens.
- Simulation is performed considering that the majority of energy transfers are made from North to South in the Central American power system (W.C. Flores, 2016).
- Simulation takes a time of $t = tsimul$.

3.2 Flowchart of the RAS

According to the operating premises, the proposed RAS flowchart is shown in Fig. 3.2.

In view of a fault in the south of the Central American power grid, the variables to be monitored are the maximum values of *Pmx1*, *Pmx2*, *Pmx3*, *t1*, and *t2*, described in the previous section; under a total simulation time of *tsimul*. These values are the inputs of the flowchart shown in Fig. 3.2. The values of the links between the countries PGU-ES: Guatemala-El Salvador and PHN-NI: Honduras-Nicaragua, as well as the link of a Nicaragua line, PLine-NI, will oscillate between different values during the simulation. These values are shown in Fig. 3.2.

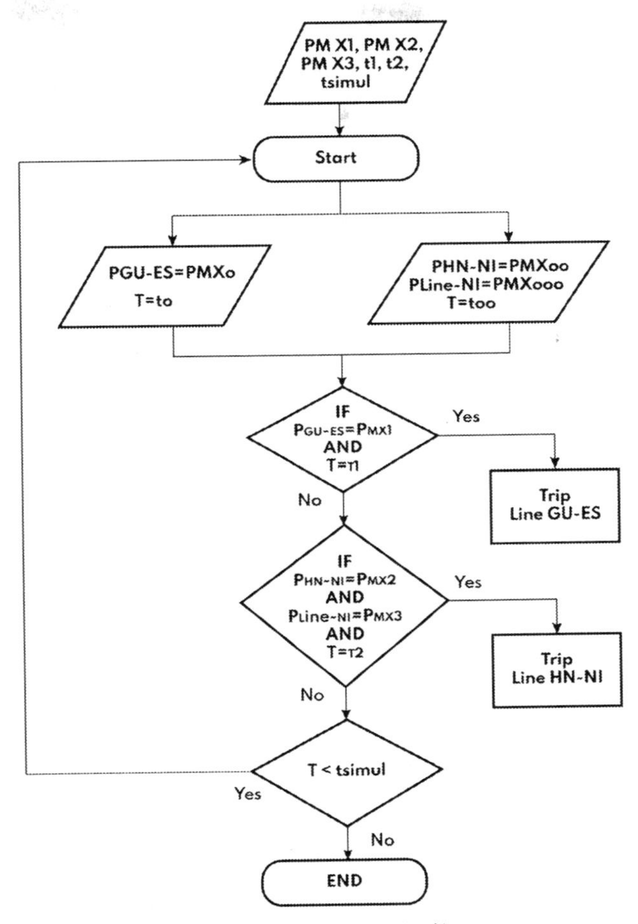

Fig. 3.2 Flowchart of the RAS. *(No permission required.)*

If the PGU-ES link reaches the value of *Pmx1* at a time $T=t1$, the lines between Guatemala and El Salvador are tripped, which will form two islands as follows: Mexico-Guatemala and the rest of Central America (see Fig. 3.1). Both islands must be stable in their operation after the failure. If those values of power and time are not reached, nothing happens and the simulation continues while $T < tsimul$.

Just the same, if the PHN-NI link reaches the value of *Pmx2* and in addition the line PLine-NI reaches the value of *Pmx3*, at a time $T = t2$, the lines between Honduras and Nicaragua are tripped, which will form two islands, as follows: Mexico-Guatemala-El Salvador-Honduras and the rest of Central America (Fig. 3.1). Both islands must be stable in their operation after the failure. If those values of power and time are not reached, nothing happens and the simulation continues while $T < tsimul$.

Trips of lines are coordinated in such a way that they cannot form more than two islands.

4. Simulation

Although a fault can occur in any site of the network, four representative cases of large power generation outage are analyzed, to show the RAS effectiveness under failures located in the south of the transmission grid. To make this analysis, a test power grid is used.

For each outage case, three base scenarios were analyzed, corresponding to three demand levels (max, rest, and valley). To obtain the four cases of analysis, many failures that cause imbalances in the generation—demand levels of the region and that force greater transfers through the interconnection lines between countries were analyzed. The sequence of events includes a three-phase short-circuit in a high-voltage bar associated with a generating power plant, the subsequent fault clearance, and the outage of at least one generator of the plant involved.

For Figs. 3.3, 3.5, and 3.7, the colors have the following meaning: *Violet* (*dark gray* in print version): power flow in line of Nicaragua (Pline-Ni); *Brown* (*gray* in print version): power flow in Mexico-Guatemala link; *Red* (*light gray* in print version): power flow in line Guatemala-El Salvador (line 1); *Green* (*dark gray* in print version): power flow in line Honduras-Nicaragua, and *Blue* (*gray* in print version): power flow in line Guatemala-El Salvador (line 2). In the same way, colors from Figs. 3.4, 3.6, 3.8, and 3.10 are *Red* (*light gray* in print version) for frequency in Guatemala and *Blue* (*gray* in print version) for Panama.

4.1 Case #1: Outage of power generation in Costa Rica

The outage of a generation power plant composed by three generators in Costa Rica (180 MW) triggered the RAS actuation due to the transitory increase of the power flows through the Honduras-Nicaragua interconnection. This event triggers the formation of two subsystems, as follows: The South, made up of Nicaragua, Costa Rica, and Panama and the north, which includes Mexico, Guatemala, El Salvador, and Honduras.

Fig. 3.3 shows the situation before and after the operation of the RAS. It acts when the power flow through a line in Nicaragua exceeds a maximum value of operation during a certain time.

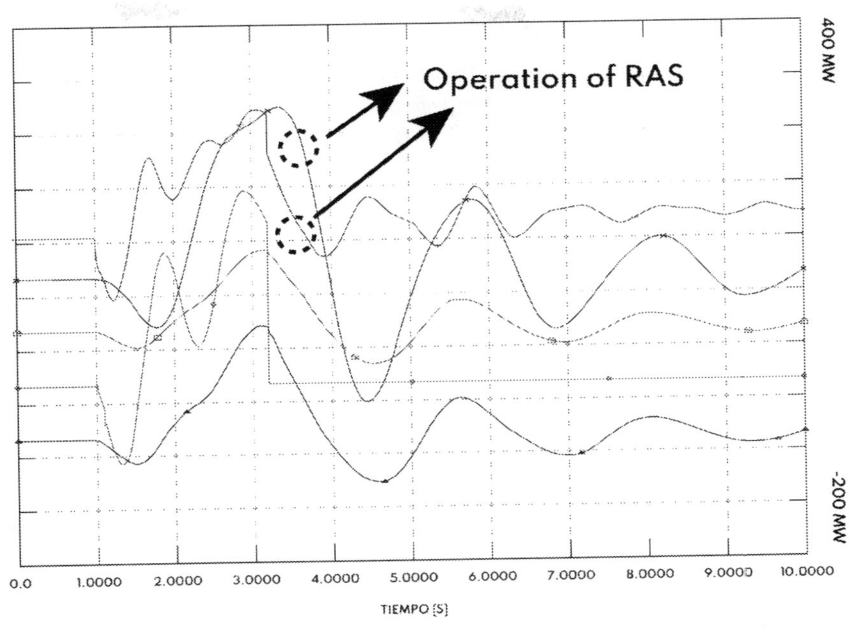

Fig. 3.3 Power flow through some interconnection lines due to the trip of three generators in Costa Rica (180 MW). *(No permission required.)*

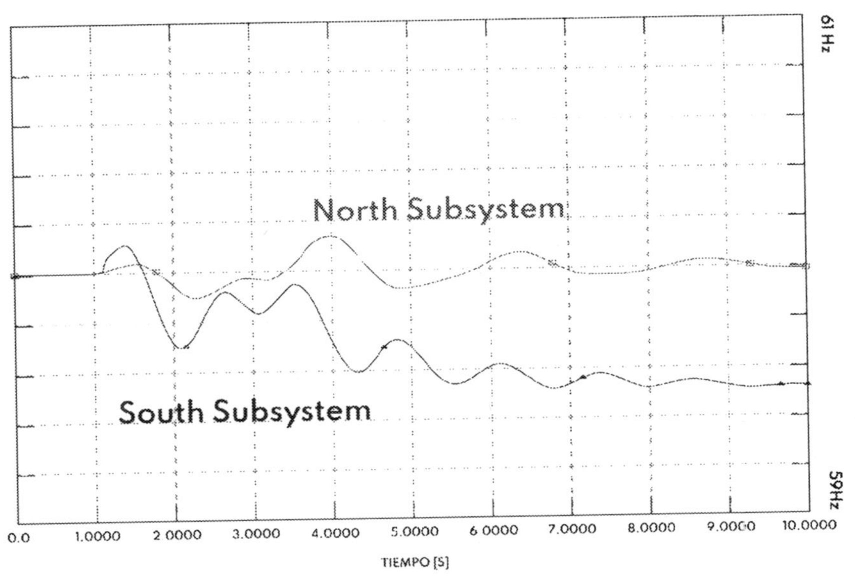

Fig. 3.4 Frequency of the two subsystems due to the outage of three generators in Costa Rica (180 MW). *(No permission required.)*

As a result of the RAS operation, the south subsystem undergoes a decrease in frequency, although it is stabilized by the operation of primary reserves (Fig. 3.4). Therefore, as a conclusion, in this case, the operation of the RAS contributes to the stabilization of the system under failures of important generation power plants located in the electric power system of Central America.

4.2 Case #2: Outage of power generation in Panama

The outage of three generators of a hydroelectric power plant (which contributed to the 210 MW in the condition before failure) causes an increase of the transfers in the north-south direction. Hence, the limits of the RAS are exceeded temporarily, triggering the opening of the interconnection lines from Honduras with Nicaragua (Fig. 3.5). In this way, two islands are formed.

Subsequently, the two subsystems are stabilized because of the action of the primary reserves. The frequency in the south subsystem drops to 59.3 Hz (Fig. 3.6).

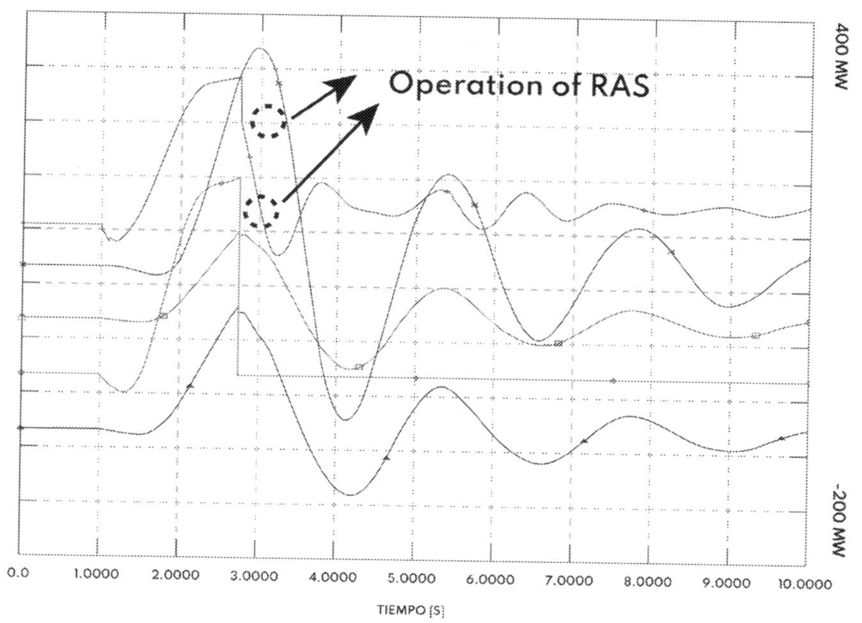

Fig. 3.5 Power flow through some interconnection lines due to the trip of power generators in Panama (210 MW). *(No permission required.)*

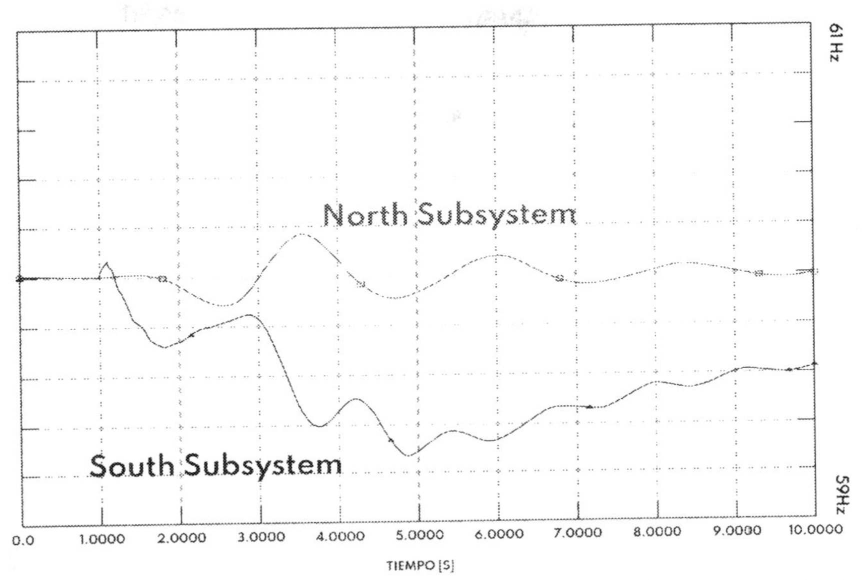

Fig. 3.6 Frequency of the two subsystems due to the outage of power generation in Panama (210 MW). *(No permission required.)*

4.3 Case #3: Outage of power generation in El Salvador

The failure and subsequent outage of 2 units of 90 MW each located in El Salvador causes an increase in imports of the power flows through the Mexico-Guatemala and Guatemala-El Salvador interconnections. However, the limits of the RAS are not exceeded (Fig. 3.7).

Consequently, whole of the Central American power system is kept in synchronism, which is evidenced in Fig. 3.8.

4.4 Case #4: Outage of power generation in Honduras

In this case, the base scenario was modified by increasing energy imports from Honduras, particularly a power flow between El Salvador and Honduras of 140 MW. On this scenario, a symmetrical three-phase fault was simulated, with a subsequent disengagement of a Hydroelectric Plant located in Honduras, which was generating 205 MW.

The disconnection of generation causes the actuation of the RAS because of the transitory increase of the power flows through the Guatemala-El Salvador interconnection. This event triggers the formation of two subsystems: The South, made up of El Salvador, Honduras, Nicaragua, Panama, and Costa Rica and the North, which includes only Guatemala and Mexico. Figs. 3.9 and 3.10 show the situation after the

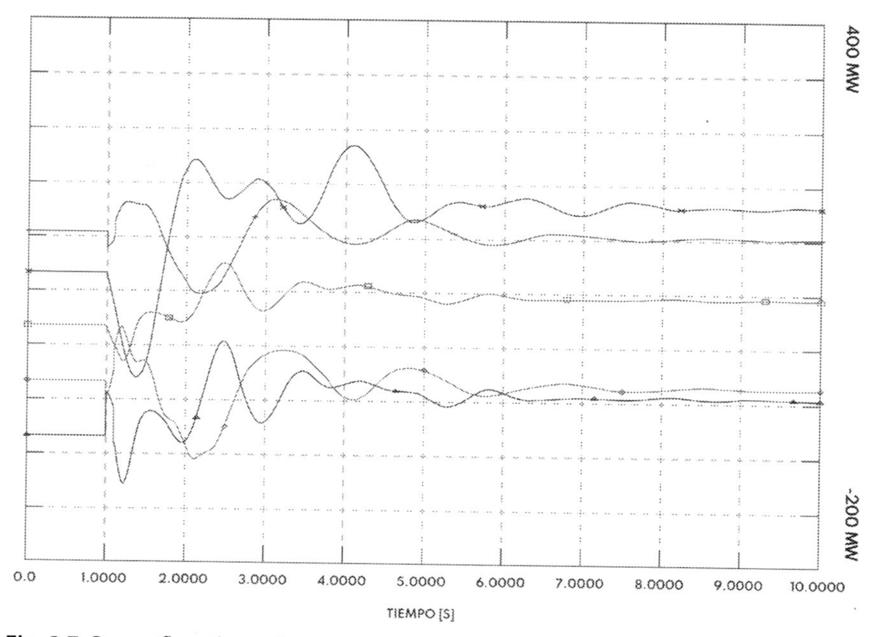

Fig. 3.7 Power flow through some interconnection lines due to the trip of power generators in El Salvador (180 MW). *(No permission required.)*

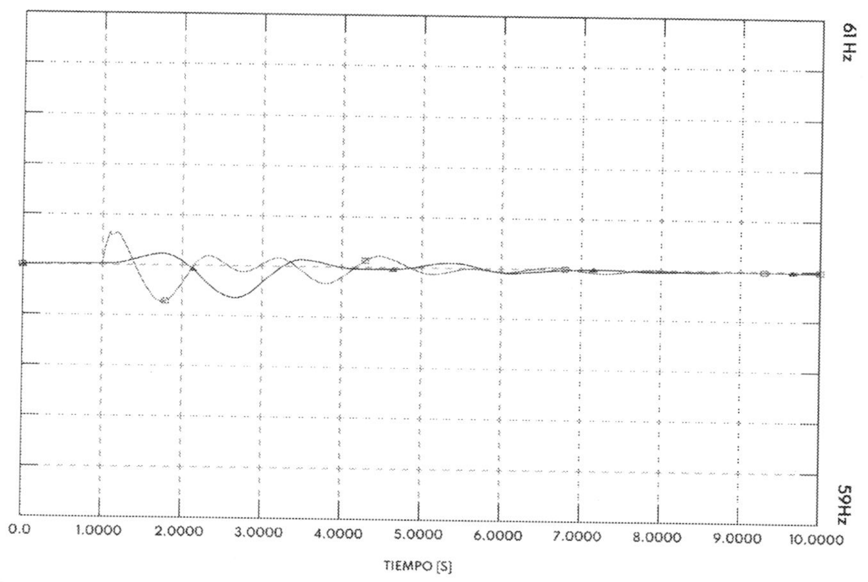

Fig. 3.8 Frequency of two points located in El Salvador (*red* (*light gray* in print version)) and Panama (*blue* (*gray* in print version)) due to the outage of power generation in El Salvador (180 MW). *(No permission required.)*

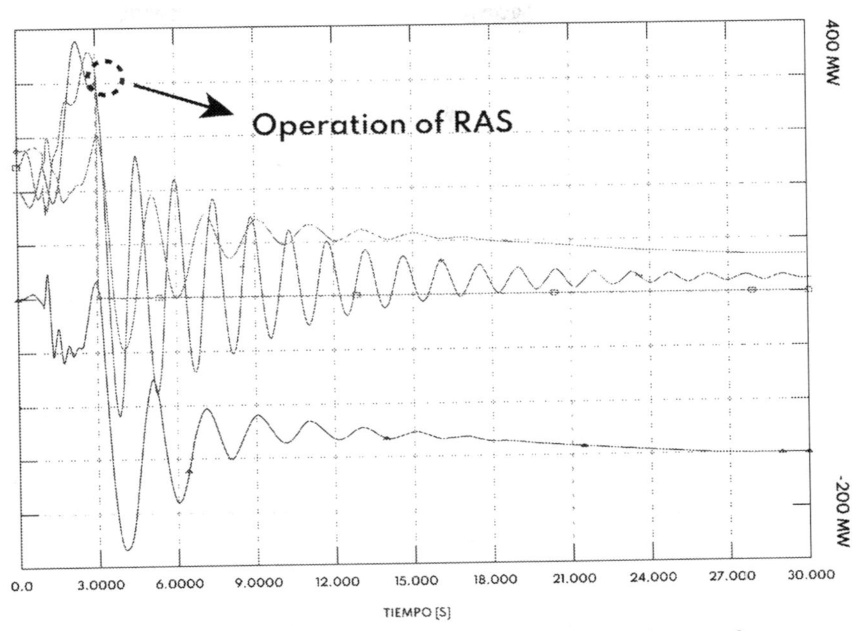

Fig. 3.9 Power flow through some interconnection lines due to the trip of power generators in Honduras (205 MW). *(No permission required.)*

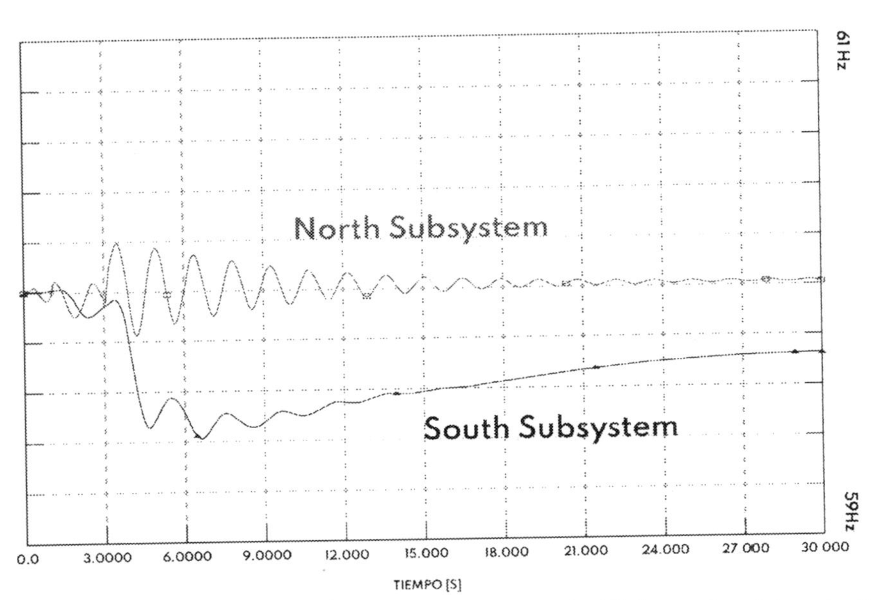

Fig. 3.10 Frequency of two points located in El Salvador (*red* (*light gray* in print version)) and Panama (*blue* (*gray* in print version)) due to the outage of power generation in Honduras (205 MW). *(No permission required.)*

operation of the RAS. It acts when the flow through the Guatemala-El Salvador interconnection exceeds 245 MW for 0.6 seconds.

Fig. 3.9 shows important fluctuations in the interconnections between countries in which the unacceptable magnitude in the case of the links between Mexico-Guatemala (violet (dark gray in print version)), Honduras-Nicaragua (blue (gray in print version)), and Nicaragua-Costa Rica (green (dark gray in print version)) stand out.

As a result of the operation of the RAS, the south subsystem undergoes a decrease in the frequency. Although it is stabilized by the operation of the primary reserves, a large time is observed to obtain a stable condition.

5. Conclusions

The use of remedial action schemes (RAS) in an electrical power system improves the performance of the electrical network. This work demonstrates that the use of an RAS in a test power grid of Central America increases the resiliency of the system under serious failures in the network. Without the appropriate regional protections, some of these faults could cause major repercussions, such as a regional blackout. By simulating four fault conditions in the test power grid of Central America, it has been shown that the operation of the electrical power system is kept stable by the use of an RAS programmed in such a way that, in spite of created islands in the electric network, these operate under stable conditions. With the implementation of this RAS, customers and users of the Central American Power Market could be trustful that the power system will keep operating even under extreme conditions. The use of RAS contributes to the stabilization of the power system under failure and large loss of power generation.

References

AETS, BCEOM and EDE Ingenieros. (2010). *Proyecto GAUREE 2, Estrategia de ENEE en Uso Racional de la Energía y Manejo de Demanda en los Sectores Industrialy Comercial, MODULO 4*. EuropeAid, Comisión Europea. Rep. EuropeAid/117195/C/SV/HN.

Flores, W. C. (2012). Analysis of regulatory framework of electric power market in Honduras: Promising and essential changes. *Utilities Policy*, *20*(1), 46–51. https://doi.org/10.1016/j.jup.2011.11.006.

Flores, W. C. (2016). "Some issues related to the regulatory framework and organizational structure of the central american electricity market," 2016 IEEE PES Transmission & Distribution Conference and Exposition-Latin America (PES T&D-LA), 2016, pp. 1-5, doi: 10.1109/TDC-LA.2016.7805635. In IEEE (Ed.) (pp. 1–5). IEEE.

Flores, W. C., Barrionuevo, J., Atlas, E., & Torres, S. P. (2017). An approach based on remedial action scheme to increase resiliency under failures in the Central American power grid. In *2017 IEEE PES innovative smart grid technologies conference-Latin America, ISGT*

Latin America 2017, Vols. 2017 (pp. 1–6). Institute of Electrical and Electronics Engineers Inc. https://doi.org/10.1109/ISGT-LA.2017.8126725.

Flores, W., Hernandez, J., & Castellón, R. (2012). El Ente Operador Regional: Nuevas aplicaciones en la operación de la Red Regional Centroamericana. In *Proceedings of 2012 IEEE CONCAPAN XXXII*.

Gómez-Expósito, A., et al. (2009). *Electric energy systems. Analysis and operation. Vol. I* (p. 119). CRC Press.

Grigsby, L. L. (2001). *The electric power engineering handbook* (pp. 11–24). CRC Press and IEEE Press.

Kundur, P., Paserba, J., Ajjarapu, V., Andersson, G., Bose, A., Canizares, C., Hatziargyriou, N., Hill, D., Stankovic, A., Taylor, C., Van Cursem, T., & Vittal, V. (2004). Definition and classification of power system stability. *IEEE Transactions on Power Systems, 19*(3), 1387–1401. https://doi.org/10.1109/TPWRS.2004.825981.

Martinez, E., & Messina, A. R. (2008). "Modal analysis of measured inter-area oscillations in the Mexican Interconnected System: The July 31, 2008 event," 2011 IEEE Power and Energy Society General Meeting, 2011, pp. 1-8, doi: 10.1109/PES.2011.6039383. In IEEE (Ed.), IEEE.

Martinez, E. M., Vanfretti, L., & Sevilla, F. R. S. (2013). Automatic triggering of the interconnection between Mexico and central America using discrete control schemes. In *2013 4th IEEE/PES innovative smart grid technologies Europe, ISGT Europe 2013*. https://doi.org/10.1109/ISGTEurope.2013.6695333.

McCalley, J. (2010). *System protection schemes: Limitations, risks, and management*. https://pserc.wisc.edu/documents/publications/.../S-35_Final-Report_Dec-2010.pdf.

Rueda, J. L., & Shewarega, F. (2009). Small signal stability of power systems with large scale wind power integration. In *XIII ERIAC CIGRE, C1, 24–28 May, 2009*.

Sanchez, G. A., Pal, A., Centeno, V. A., & Flores, W. (2011). PMU placement for the Central American power network and its possible impacts. In *IEEE PES Conference on Innovative Smart Grid Technologies Latin America* (pp. 1–7). IEEE. 10.1109/ISGT-LA.2001.6083202.

U.S. Department of Energy. (2014). *Smart grid investments improve grid reliability, resilience, and storm responses*. https://energy.gov/sites/prod/files/2014/12/f19/SG-ImprovesRestoration-Nov2014.pdf.

Vance, K., Pal, A., & Thorp, J. S. (2012). A robust control technique for damping inter-area oscillations. In IEEE (Ed.), *2012 IEEE Power and Energy Conference at Illinois* (pp. 1–8). IEEE. 10.1109/PECI.2012.6184591.

WECC. (1996). *Remedial action scheme design guide*. https://www.wecc.biz/Reliability/Remedial%20Action%20Scheme%20Design%20Guide.pdf.

CHAPTER FOUR

Protective relay resiliency in an electric power transmission system

Arturo Conde Enríquez and Yendry González Cardoso
Universidad Autónoma de Nuevo León, San Nicolás de los Garza, Nuevo León, Mexico

1. Introduction

Power system protection has always been a vital aspect of electrical power system engineering. Its main objective is to maintain a secure and reliable network by isolating an element or a faulted section of the network during any adverse condition that may occur during operation. A safe and dependable protection system helps to reduce equipment damage, reduce power interruptions, improve power quality, and, most importantly, improve the electrical system safety (Dehghanian et al., 2019).

Security and reliability, to some extent, can be achieved with the help of protective devices installed throughout the network. Overcurrent relays (OCR) are the earliest adapted protection scheme and currently widely used due to their simple operation characteristics, reliability, security, dependability, and low-cost, which are basic functional requirements of any network protection (Coffele et al., 2015). When used on interconnected systems, these relays should include a directional discriminating feature; the directional function of the overcurrent relay provides selectivity to interconnected systems. Directional overcurrent relays (DOCRs) are capable of discriminating between the forward and reverse direction current on its protected line, which is an important feature understanding that interconnected networks with multiple sources can contribute to fault current on either side of the protected line, causing an unnecessary trip on the system whether the directional function is enabled or not. Protection engineers have taken on the arduous task of looking for alternative methods that can assist in improving the operation times of these relays so that they can provide better sensitivity for fault conditions on looped systems (Huchel & Zeineldin, 2016).

Electric Power Systems Resiliency
https://doi.org/10.1016/B978-0-323-85536-5.00006-0

Overcurrent relays have a highly dynamic protection zone as this is more affected by the operating conditions of the electrical network, such as topological changes, normal demand variation, and operating states that modify the fault current contributions. However, due to its functional simplicity, whereby it responds appropriately according to the fault conditions of the system, and has a tolerance of temporary overloads, its application occurs at all voltage levels of the electrical network. The adjustment process is complex as the coordination requirements present a functional dependency with other overcurrent relays. Thus, the operating time is determined by selecting a time curve that guarantees coordination; this is the only protection principle that presents this behavior. Relay coordination is a complex process due to the number of elements and the resulting coordination pairs; this is proportional to the electrical network interconnection. It is advisable to formulate coordination as an optimization problem (Ezzeddine & Kaczmarek, 2008).

The application of DOCRs has sensitivity limitations and may present high operation times for electrical system protection. Operating under overload conditions combined with minimal fault magnitudes of distributed energy resources (DERs) under weak interconnection can compromise fault detection. To avoid false operations, it is common practice that the setting values be determined with the maximum values of the load current, subject to contingencies $n-1$; this can result in settings that compromise the sensitivity for fault detection. Furthermore, minimal fault currents can be very common when the utility contribution is small or in isolated operation. Additionally, the connection with utility systems can create similar fault currents on all buses (Yousaf et al., 2018).

1.1 Impacts of DG penetration on protective relay coordination

Electric grids have evolved in their topology, transforming from centralized to decentralized systems due to the interconnection of distributed generation (DG), specifically of renewable energy sources (RES). These changes in networks were caused by the increase in demand in recent decades and deregulated energy policies. These electrical grids mostly operate in an interconnected manner to the utility and sometimes operate in isolation when a failure occurs in the system, or even when a microgrid is weakly connected to the utility (Razavi et al., 2019).

Renewable energy has become one more option to support electrical grids, becoming increasingly attractive. However, the intermittency and uncertainty of these types of sources pose significant challenges in achieving deeper penetration into existing power systems. Another characteristic of intermittent renewable sources in the networks is the lack of inertia during high penetrations of these sources; the inertia of the power system is decreased due the reduction in rotating machine interconnections. As a result, electrical systems will require the adaptation of new strategies or operational alternatives to include these sources (Peças Lopes et al., 2007).

Despite the numerous advantages of having DGs installed in the network, there are also negative impacts on the protective relays. The design of the low-voltage electrical system has traditionally been designed to operate radially, so the protection schemes are generally shaped by overcurrent protections. These protection schemes are traditionally adjusted to maximum border values to ensure dependable and safe operation; the relay settings are constants without actualization by the operational state of the power grid. The topological changes in interconnected configurations with bidirectional flows, and the highly intermittent operation, mean that conventional schemes hardly meet the protection requirements (Singh, 2017).

The presence of renewable energy sources (RES) affects the performance of the protection in two ways; in the case of small conventional sources such as turbo diesel, it causes an increase and change in direction of fault currents in areas of the circuit where there are no contemplated directional schemes. The direct impacts on the protection system are false tripping, under reach or sensitivity loss of relays, and coordination loss between primary and backup relays. On the other hand, in case of being RES, the high Thevenin impedance causes a limited fault contribution in both magnitude and time; in the worst case, it maintains a limited contribution to the fault avoiding successful breaker reclose. In the case of isolated microgrids or with a high-impedance connection in the utility, the reduction in fault values compromises the sensitivity of the protection scheme, leading to faults that are difficult to detect. In the case of small MGs connected to the utility, the fault current profiles are similar in all nodes, showing difficulty coordinating protections (Memon & Kauhaniemi, 2015).

1.2 Small conventional sources

The connection of small controllable generators such as turbodiesels, or battery sources, results in an increase in the fault current that can affect relay coordination. The operation time between relays is reduced, and it can be so close that the selectivity can be lost, with simultaneous trips (sympathy trips) (Mladenovic & Azadvar, 2010).

The loss of coordination results from the increase in the fault current values that were initially considered; when this value is increased by small sources, the operating times of both relays are reduced because the time curves are convergent, and the interval Coordination time (*CTI*) between relays is lost; this results in a sequential operation between relays affecting selectivity by the tripping of healthy lines.

Fig. 4.1 shows the effect of DG contribution in relay coordination. As it is a radial network, the fault current from the utility is considered; the goal is to have an operation that guarantees the release of the fault. Thus, for the fault F_1, the relay R_1 must operate as primary protection; in case the interruption of the circuit is not achieved, the relay R_2 will operate as backup protection, guaranteeing an operation time difference of no less than *CTI* (commonly 0.2–0.3 s). With this sequence, the dependability of the system is increased, guaranteeing the removal of the fault. The operation of the relay

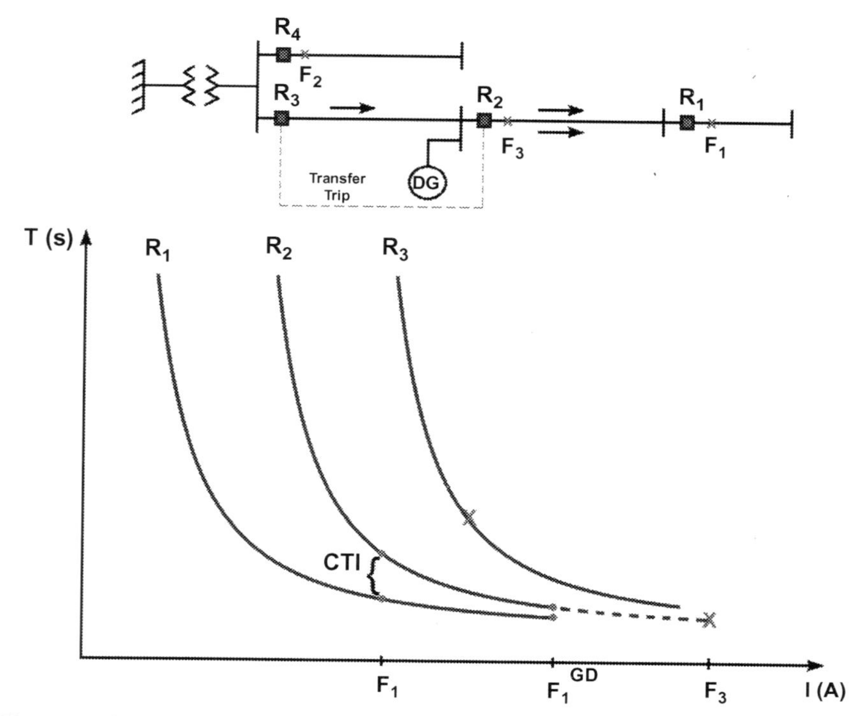

Fig. 4.1 False tripping due to DG penetration.

R_3 as a backup-backup protection is not required because the malfunction of protection in two different substations for the same event is very unlikely.

With the additional contribution of DG, it is possible that the time of the relays decreases, and simultaneous trips occur due to lower *CTI*; this is shown in the dashed lines in Fig. 4.1 (F_1^{GD}). This contribution may cause a malfunction in other feeders as the trip of the Relay R_3 for the fault F_2. To solve this problem, it is necessary to update the fault currents in such a way that the additional contribution of the generator is considered. In the case of small generators, it is common for them to be used for a few hours of the day, for example, to reduce the maximum demand. However, the setting relay must be made considering the maximum fault current. When the GD is not connected, the operating times of the relays will be longer. It is feasible to have discrete adaptive schemes with specified relay setting groups depending on the GD connection, or in a continuous adaptive scheme to have re-coordination schemes on-line by a protection system that monitors the GD contribution.

On the other hand, it is necessary to consider both the magnitude of the current increase and its effect on the relay coordination. If the current increase is very small, the time reduction will be too, so it may not be a coordination problem. On the other hand, depending on the relay setting, the time curve may be in its time asymptote; in this horizontal region of the curve, the variation of time may be very small even when the GD current contribution is large without losing coordination. This can be observed in relay R_1; its operating time remains relatively constant even with increasing current. In case of the relay R_2, which is also in its time asymptote, the *CTI* can be kept within safe values without affecting the coordination. Another effect of the GD contribution is the infeed that it causes between relay R_2 and relay R_3 for fault F_2, as observed in Fig. 4.1, where the increase in current accelerates R_2, but R_1 does not see this increased current so is not accelerated; the *CTI* is higher but the coordination sequence is maintained. In case the operating times of relay R_3 are not permissible, a trip transferred to relay R_3 can be enabled when the fault is detected in relay R_2.

1.3 Connection of RES

The connection of small networks is beneficial because the reactive power deficiencies will be provided by the utility maintaining a voltage control; thus, the energy imbalances in the microgrid (MG) will be supplied or absorbed (Fig. 4.2A).

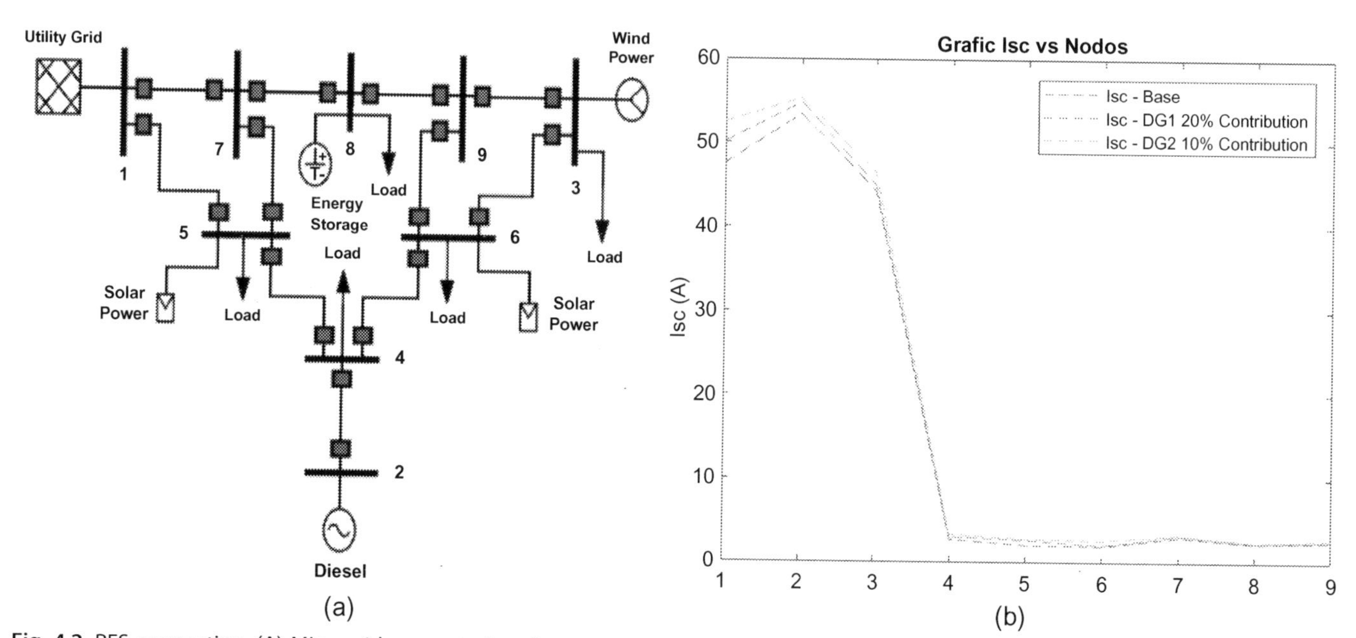

Fig. 4.2 RES connection. (A) Microgrid connected to the power grid, (B) Fault current profiles.

However, as a result of the strong connection, the fault profiles will be flat because the fault contribution is mainly contributed by the utility, as the RES contribution is limited in magnitude and time, and due to the low impedance of the MG (Peterson et al., 2019). The resulting fault current profiles for the analyzed system are shown in Fig. 4.2B.

Depending on the interconnection of the system, the complexity in coordination for radial systems will result in high tripping times because the coordination will be performed for similar magnitudes of fault currents (Fig. 4.3). In meshed systems, the coordination obtained will be unfeasible in many scenarios due to noncompliance with the coordination restrictions. Selective fault detection will be difficult and incorrect trips can occur with more relays tripping than necessary.

1.4 Islanding operation

As a result of the disconnection of a section of the electrical network due to the presence of a fault or abnormal operating conditions, the isolated section may be in an unintended island mode (Chowdhury et al., 2008). It is common for the electrical island to present energy imbalances that are not sustainable, and it is necessary to enable disconnection schemes for generation sources and loads on the island to avoid damage (IEEE 1547-2018, 2018). In another sense, there are schemes reported to save the island based on finding a sustainable balance through the management of energy resources during

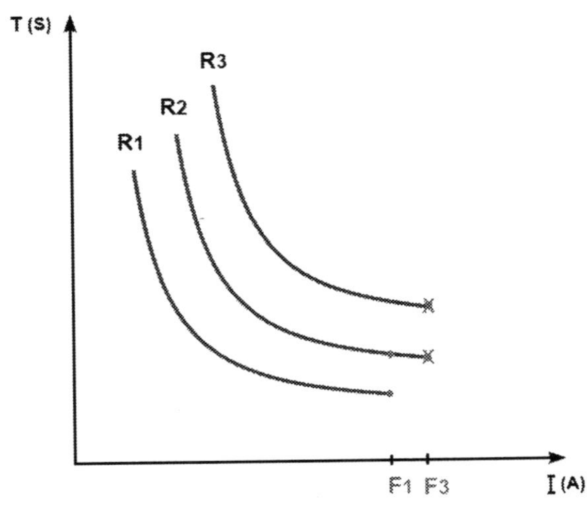

Fig. 4.3 Fault detection with RES connection, coordination with similar magnitudes of fault current.

the time of operation on the island until reconnection with the utility is possible. In networks with a high presence of RES, the operating time during islanding conditions can be short due to the intermittency of the sources and the lack of dispatchability. In cases where there is controllable energy support, such as a turbodiesel or battery bank, the operating times on the island can be increased; to achieve this, energy resource management systems (DMS) may be required. In Fig. 4.4, the performance of MG under dynamic conditions for RES for 24 h is evaluated to compute the frequency and voltage profiles (Pérez et al., 2019). The sensitivity of the protections must be determined to ensure the dependability of the protection scheme, since it is very common for the protections to have insufficient sensitivity for fault detection because the RES that commonly present on the island have low fault currents and, in many cases, to have high operating times.

On-line protection schemes have traditionally been limited in transmission systems due to the technical difficulty of implementing control actions toward the electricity grid. Adaptable load shedding schemes and dynamic electrical islands for out-of-step stability have not been possible to implement. However, these technical limitations are generally not a problem because MGs are small networks, and the monitoring and the communication systems are integrated because it is economically feasible to implement. This can lead to having systems on-line that can adjust settings such as the pickup current being calculated with the current values of the load current, thus increasing the sensitivity. Protection schemes can change their configuration parameters and coordination in response to topological changes in the network. These schemes can be implemented in time periods with dynamics similar to the behavior of the demand, so update times of 5–10 min can be feasibly carried out without degrading the protection functions. In the case of topological changes, they can be activated in an updated emerging system.

2. Benchmark of overcurrent coordination

In recent years, overcurrent relays (OCRs) have been the most reliable protection schemes applied to electrical networks worldwide due to their simple operation principle and the fact that they comply with the core objectives of system protection reliability, selectivity, speed, simplicity, and economics (Bedekar et al., 2011). These qualities make OCRs the best option for protection engineers to select from the available types of protection schemes for both distribution and transmission networks. The operation

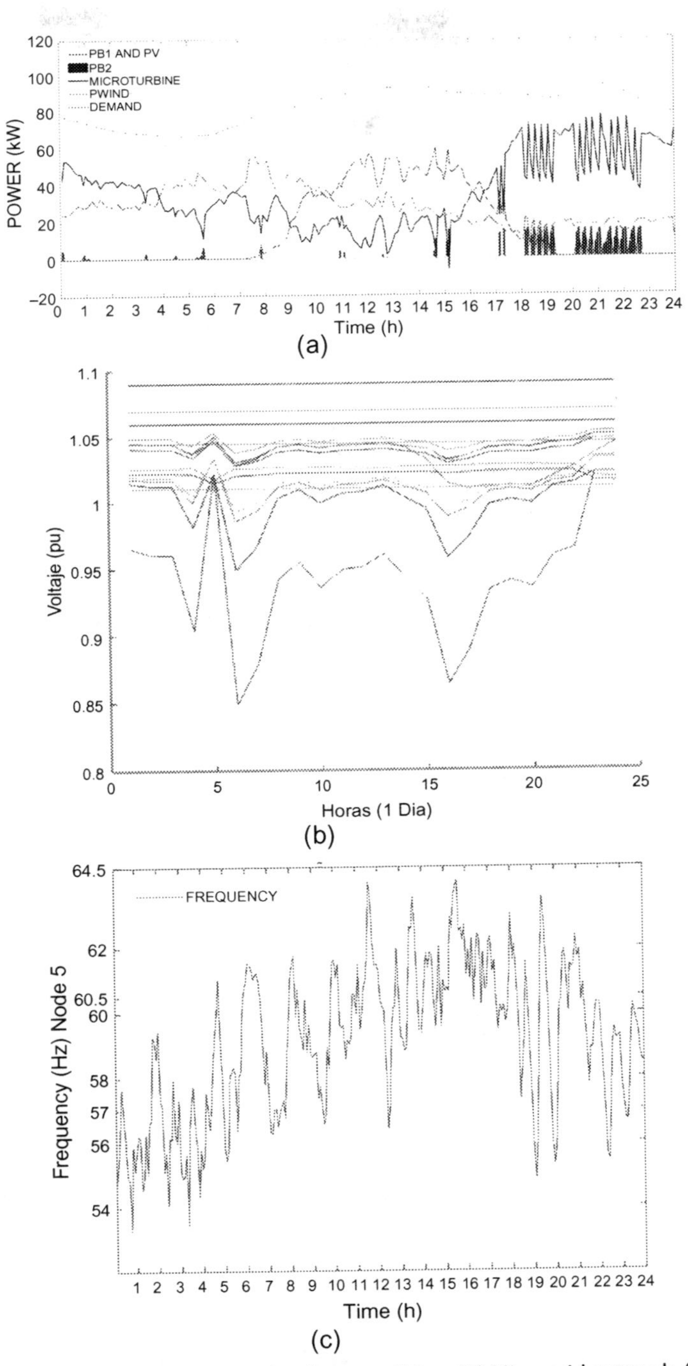

Fig. 4.4 Operation of MG during islanding condition. (A) Microgrid network, (B) voltage profiles, (C) frequency response.

of the relay is triggered when the measured current causes a direction control unit and overcurrent unit to operate after its magnitude surpasses the assigned pickup current. The correct operation of the relay is ensured by setting a pickup current that is greater than the maximum possible load current (I_{Load}) but below the minimum fault current that can be seen on the protected line by using a plug setting multiplier (*PSM*):

$$I_{pickup} = I_{load} * PSM \tag{4.1}$$

The *PSM* is a security factor that is used to prevent any unnecessary trips caused by line overload and current measurement errors. *PSM* values are normally between 1.4 and 2, which should ensure the reliability of the relay operation and at the same time provide sufficient sensitivity to detect minimum fault currents at the adjacent far line-end under minimum generation.

Inverse time relays operate in less time for higher fault current magnitudes and vice versa for smaller fault current magnitudes. The high selectivity ensures that relays closest to the fault operate first, while upstream relays function as backup protection; this is known as selective coordination. The time-current characteristic curves give the classification of this type or relays (4.2). The time curve model is from either IEEE C37.112 (C37.112-2018, 2019) or IEC 60255-151 (IEC 60255-151:2009, 2009); the different curves are obtained by means of constants presented in Table 4.1.

$$t_{IEEE} = \left[\frac{A}{\left(\frac{I_{sc^{3\varnothing}_{max}}}{I_{pickup}} \right)^{p} - 1} + B \right] * TDS,$$

$$t_{IEC} = \left[\frac{A}{\left(\frac{I_{sc^{3\varnothing}_{max}}}{I_{pickup}} \right)^{p} - 1} \right] * TDS \tag{4.2}$$

where:

t = relay operation time; $I_{sc^{3\varnothing}_{max}}$ = maximum three phase short circuit current

I_{pickup} = pickup current setting; A, B, p = constants of IEEE and IEC standards; TDS = Time dial setting

The time-dial setting (*TDS*) o *dial* represents the journey of the disk (integral of the velocity with respect to time) in electromagnetic relays. A comparison of the different standardized curves is shown in Fig. 4.5A for IEEE/ANSI

Table 4.1 Constants of IEEE and IEC standard relay time curves.

Standard	Curve type	*A*	*B*	*p*
IEEE-C37.112	MI—Moderately Inverse	0.0515	0.1140	0.02
	VI—Very Inverse	19.61	0.491	2.0
	EI—Extremely Inverse	28.2	0.1217	2.0
	NI—Normally Inverse	5.95	0.18	2.0
	STI—Short-time Inverse	0.02394	0.01694	0.02
IEC-60255	SI—Standard Inverse (C1)	0.14	0	0.02
	VI—Very Inverse (C2)	13.5	0	1
	EI—Extremely Inverse (C3)	80	0	2
	STI—Short-time Inverse (C4)	0.05	0	0.04
	LTI—Long-time Inverse (C5)	120	0	1

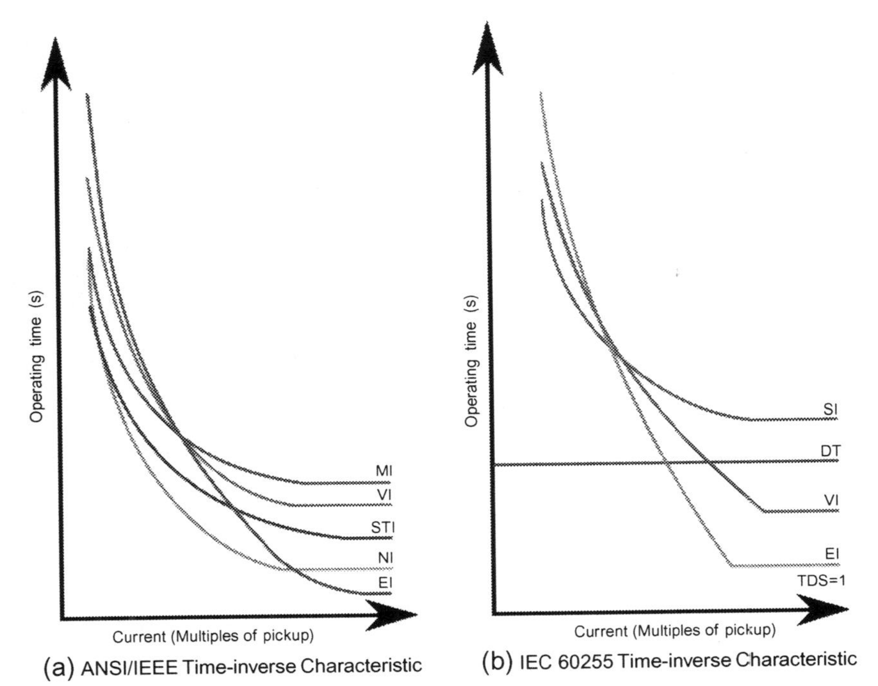

Fig. 4.5 Time-current tripping characteristic of DOCRs.

curves and Fig. 4.5B for IEC using a common setting current and TDS setting of 1.0.

When setting the DOCRs that the type of curve must accomplish with the requirements of protection function, the use of very inverse curves is common for transmission and distribution systems because they provide an adequate tolerance to temporary overloads and coordinate well with transformer curves. On the other hand, for industrial systems, the coordination is very complicated because DOCRs must coordinate with other overcurrent protections such as fuses and low-voltage switches, in addition to equipment damage curves; therefore, the use of different types of DOCR curves is very common.

The inversion grade is given by the characteristic curve constants (*A*, *B*, *p*), while the *TDS* determines the time multiplier setting used to increase or decrease the time taken to obtain coordination between relays; this is the goal of the coordination process. *PSM* determines the vertical asymptote of a curve, as shown in Fig. 4.6. Here, these parameters are used like grades of liberty in coordination algorithms due to the proportionate possible solution for obtaining relay coordination.

The sensitivity analysis evaluates the feasibility of the backup function of each relay; for that, relays should have the ability to detect minimum fault currents that occur at the far end of the protected line. Failing to reach sensitivity means that relays have either high or unacceptable operating times. To overcome the sensitivity, each relay in the network is analyzed as follows:

$$S_b = \frac{Isc_{2\varnothing}}{I_{pickup}} \geq 1.5 \tag{4.3}$$

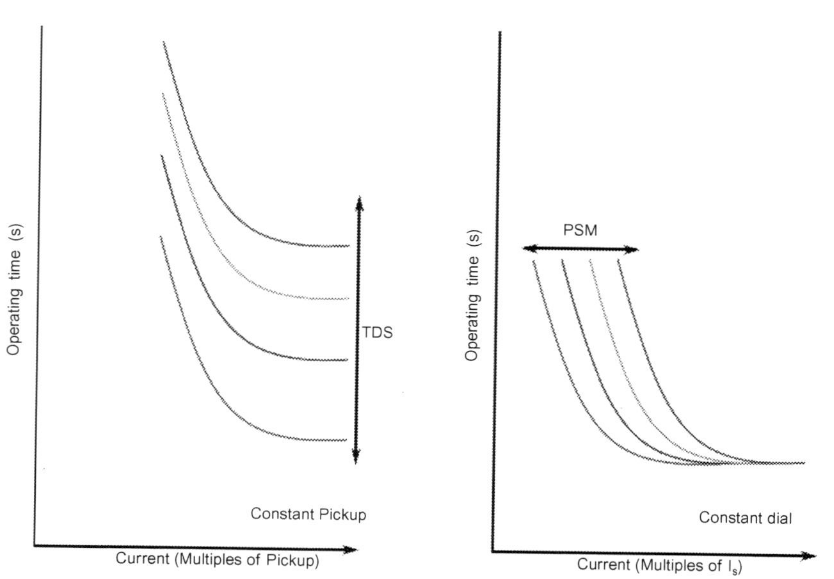

(a) Time dial multiplier setting effect on TCCs (b) Plug multiplier setting effect on TCCs

Fig. 4.6 Effect of *TDS* and *PMS* on inverse-time curves.

where $Isc_{2\varnothing}$ is the minimum fault current at the remote line-end. A lower limit of 1.5 is considered in Eq. (4.3) because the operation times are large and useless for the protection of multiple currents ($M = I_{sc}/I_{pickup}$) between 1 and 1.5; this can be considered a dead zone for protection purposes (Fig. 4.7). The curve of M versus T was generally provided for coordination; the protection engineer had to verify that the active zone of the relay is used for the adjustment. The range of M was established from 1.5 to 20, and the value of 20 was used to prevent there being more than 100 A in the secondary and damaging the relay.

2.1 Overcurrent coordination

It is important for the protection to ensure reliability; for that, the concept of coordination is introduced. Coordination between relays is necessary to achieve a desired sequence of relay operations; the primary relay should operate before any other relay to detect faults on its own line. The coordination process starts from downstream to upstream relays; the manual coordination is always done for pairs of relays. Considering the radial feeder in Fig. 4.8, for fault F, relay R_3 (primary relay) should operate to eliminate the fault current on its protected line. However, it is possible that relay R_3 is defective and did not see the faulted current, hence the need for relay R_2 (backup relay) to operate and remove it. The setting of relay R_3 is known; thus, the objective of coordination is to obtain the dial of relay R_2. The time curves of DOCRs are convergent; for fault F, the minimum interval between relays is achieved. For that, the operation times of relays should comply with a coordination time interval (CTI). Values that warrant the

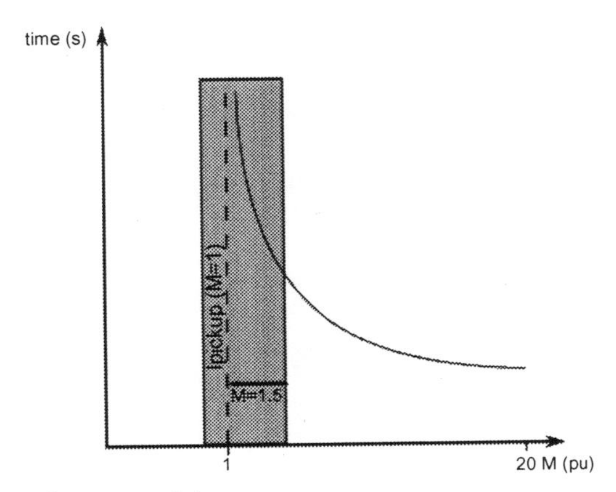

Fig. 4.7 Multiples of current of time curve.

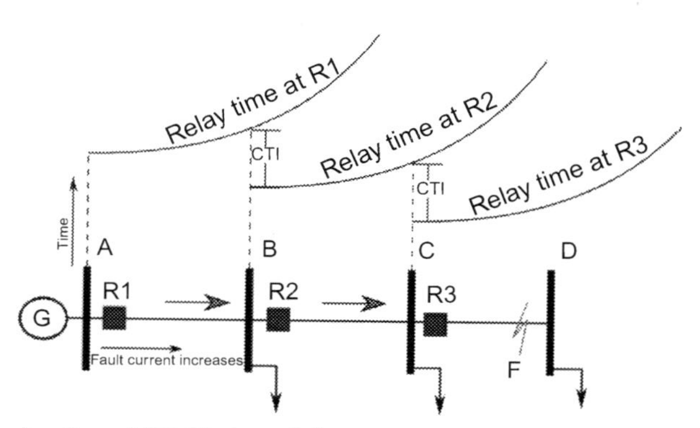

Fig. 4.8 Coordination of DOCRs in radial system.

time coordination between relays (i.e., 0.2 or 0.3) are commonly used. For the next pair coordination, the setting of R_2 is known, and this is the primary; thus, the goal is obtaining the dial of R_1 as a backup.

Considerations for coordination. In radial systems, it is common for the I_{pickup} value of backup relays to be higher than the primary relay; due to the source being at one end, the fault current profile is lower in downstream relays and higher in upstream relays; this ensures that coordination is obtained. In meshed networks the behavior is different, and it is probable that the pickup current is greater in the primary than the backup; the crossing of curves can make coordination difficult (Fig. 4.9). The overlapping of curves is given for various reasons including the use of different relay characteristic curves, different fault levels, and different relay pickup values. It is recommended that the same characteristic curve is used for radial systems to prevent this condition to some extent. However, this recommendation is not entirely applicable to interconnected systems, because it is more than likely that curve crossover will occur due to the above-mentioned reasons, in contrast to radial systems. The crossover of curves can be very deceiving for different fault levels. For higher fault levels, the coordination pair might seem to be well coordinated, but as the fault level decreases, the possibility of the backup relay operating before the primary relay is higher, causing an unwanted tripping operation. Various research have focused on trying to resolve the issue using a two- or three-point fault method evaluation, or by even including more degrees of freedom to modify and manipulate the characteristic of the relay curves (Lua & Chung, 2013).

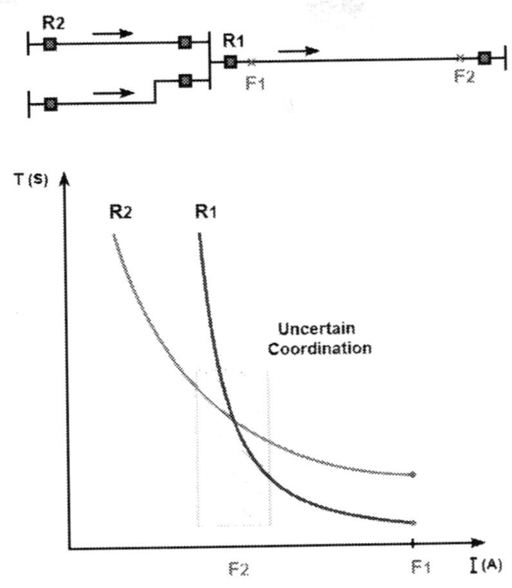

Fig. 4.9 Crossing curves in DOCRs coordination.

Another very common scenario in meshed systems is that, due to the contribution of other feeders connected to the bus, the primary relay sees more fault current than the backup (infeed); the relays have to be coordinated even when they see different currents (Fig. 4.1).

2.2 Metaheuristic optimization algorithms

The DOCRs coordination increases in complexity as the interconnected electrical system increases. The difficult task of coordinating looped networks in recent years has been solved using various analytical and graphical methods, which are the conventional solutions. The task is very tedious and challenging, with the possibility of making very drastic mistakes that can lead to undesired operations during normal or abnormal conditions of the system. Therefore, researchers have focused their efforts on trying to solve the problem using optimization techniques. Various optimization techniques have been recently explored to solve the complex coordination of DOCRs on the large interconnected system. These optimization techniques have improved the operating time of the relays to such a degree that the system can be more reliable, selective, and, more importantly, secure.

Optimization formulation. The conventional relay coordination formulated as an optimization problem seeks to minimize the relay's operation time by seeking an adequate *TDS* and *PSM*. The objective function is determined by the relay time curves, subjected to parametric and time constraints to comply with the coordination criterion. Deterministic methods (Ezzeddine & Kaczmarek, 2008; Kida & Gallego Pareja, 2016; Noghabi et al., 2010) are very challenging because they are highly dependent on a good initial guess and have a high probability of being trapped in local minima solutions. In search of the relay coordination solution, the use of heuristic optimization algorithms (Shih et al., 2015; So et al., 1997; Zeineldin et al., 2006) is very favorable due to their simplicity, flexibility, derivative-free mechanism, local optima avoidance, and robustness, showing good results in a highly restricted problem domain.

Metaheuristic algorithms use a population-based approach to solve optimization problems. The single objective formulation is very helpful in providing a straightforward solution for any optimization problem and is the most commonly used approach to solving the coordination of DOCRs. Many objective functions (OFs) have been outlined in the literature; their evaluation is not an objective of this chapter. We use the expression for the purpose of describing the coordination methodology. The weighted objective function used for the evaluation of two fault locations is given below:

$$OF = min\left\{\frac{NV}{NCP} + \omega_1\left(\frac{\sum_{p=1}^{NCP} t_p}{NCP}\right) + \omega_2\left(\frac{\sum_{b=1}^{NCP} t_b}{NCP}\right) + \omega_3\left(\sum_{pb=1}^{NCP} CTI_{pb}\right)\right\} \quad (4.4)$$

where NCP and NV are the total number of coordination pairs (CPs) and the number of CPs violated, respectively. The variables ω_1, ω_2, and ω_3 are the weighting factors used to determine the importance of each function in the OF and $CTIe_{pb}$ is a soft constrained CTI error of the pbth coordination pair. The operating time of each relay is defined as follows by taking the time curves of Eq. (4.2) as a reference.

Equality constraints. The equality restrictions are determined by the equations of time for each relay given by Eq. (4.5); the constants are previously defined. It is common for the same type of curve to be used to reduce the number of crossing curves; the fault current is determined for each coordination pair. The *TDS* and *PSM* are used in the optimization problem as degrees of freedom to obtain possible solutions (see Fig. 4.6).

$$t_{p,b} = \left[\frac{A}{\left(\frac{I_{sc3\varnothing_{max}}}{I_{load} * \boldsymbol{PSM}} \right)^{n} - 1} + B \right] * \boldsymbol{TDS} \tag{4.5}$$

Inequality constraints. The *PSM* range should ensure that the relay discriminates between normal and abnormal conditions. The *PSM* inequality hard constraint is given by Eq. (4.6). The PSM_{min} gives safety to the operation of the relay, avoiding tripping under permissible transient conditions and during temporary load conditions; the PSM_{max} avoids the need for adjustments with large values, favoring the sensitivity for fault detection and preventing the DOCR from operating with slow times.

$$PSM_{min} \leq PSM \leq PSM_{max} \tag{4.6}$$

The *TDS* is the multiplier setting used to increase or decrease the relay operation time to assist with the coordination between two relays. The maximum setting (TDS_{max}) should ensure that the resulting operation time for a fault current is not so high that it prevents damage to the network equipment. On the other hand, the minimum setting (TDS_{min}) should prevent the relay from operating as an instantaneous relay, defying the purpose of time-current grading curves. The *TDS* inequality hard constraint is given by Eq. (4.7).

$$TDS_{min} \leq TDS \leq TDS_{max} \tag{4.7}$$

It is necessary for the primary relay (t_p) to be given ample time to operate to free any abnormal conditions present on its protected line before the backup relay (t_b) operates for both near and far ended faults, thus guaranteeing the coordination between relays. The backup relay should operate only if one of the primary relays in the forward direction adjacent line fail to operate. The *CTI* inequality soft constraint is given for each one of n pairs of coordination in Eq. (4.8):

$$\left(t_b - t_p\right)_n \geq CTI \tag{4.8}$$

For an optimal solution, there should be zero miscoordination among the coordination pairs (CP) of the tested network. As the interconnected system gets larger, with the addition of lines and equipment, it is unlikely to lead to a reduction in the number of miscoordinations without expanding the range of each decision variable. The need to reduce miscoordination is a dilemma in the optimal coordination of DOCRs because it is sometimes necessary to increase the operation time of the relays to decrease the number of miscoordinations, and vice versa if the relay operation time needs to be reduced due to limitations provided by the network configuration. For good coordination, the relay must operate immediately after a fault occurs in its protected line; however, at the same time, it is necessary to have a sufficient time delay to provide backup for the adjacent line in the forward direction when the primary relay does not operate. These are related by means of *CTI* restrictions (4.7); therefore, any setting modification in one relay has a domino effect, affecting all coordination.

The weighting factors are obtained by evaluating various sets of weighting factors to obtain the pareto-front that accurately represents the effect and behavior that each objective has on the weighted objective functions. The form of the pareto solution results in nonconvex functions, so that the selection of the best solution is obtained only from the lowest value of the sum of the functions (Fig. 4.10).

2.3 Test systems

In this investigation, two test systems are used to solve the coordination problems using conventional coordination methods. In the literature, many algorithms have been recorded to solve coordination; it is not the purpose of

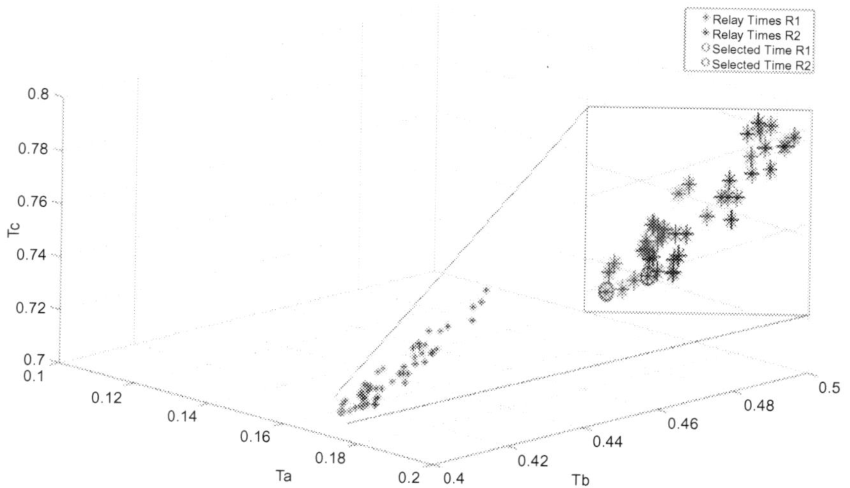

Fig. 4.10 Selection of the best solution for the relays.

this chapter to evaluate them all. The coordination of the DOCRs was carried out using a metaheuristic algorithm; the results presented show the solutions that it offers and outline proposals to improve cocrdination. The formulation of the optimization algorithm used was the Differential Evolution (DE) (Storn & Price, 1997).

The first test system is the IEEE-14 bus, the second test system is a Micro-grid, the selected voltages are 138/33/11 kV and 34.5 kV respectively. Fig. 4.11 shows the networks used in this research.

The general considerations to determine the number of relays to be used in the optimization process are described in Section 4.4. The resulting number of relays on the IEEE-14 and MG bus test networks are 30 and 16 DOCRs, respectively. These are assigned a name, using the buses that the main line is on, as described above. The fault and load currents observed by each relay in the network are provided in the Appendix section. A total of 50 and 30 coordination pairs are formed, respectively, in both test systems. Sensitivity analysis eliminated CPs that could not be coordinated due to network limitations, resulting in a total of 37 and 21 CPs in each test network, respectively.

The maximum and minimum values of PSM and TDS are in a continuous range of [1.4: 1.6] and [0.1: 3.0], respectively, to ensure that a reasonable relay operation time is obtained. The minimum CTI used in this investigation to ensure coordination is 0.2 s along the time-current curves; that is, at the maximum (start-line) and minimum fault current (end-line). The very inverse curve is used for most test cases, as it is commonly used in DOCR.

Case 1. 14-bus IEEE system under normal operation

Table 4.2 shows the resulting times for primary operation and backup obtained in the IEEE system of 14 nodes applying the conventional coordination method. From the time results presented in Table 4.2, it can be seen that there is a violation in 5 pairs of relays. The computed time to obtain results with DE is also included in the table.

Table A in the Appendix section shows the I_{sc} and I_{Load} values for this analysis. Fig. 4.12 shows the loss of coordination between one of the pairs of relays; the backup relay acts in a shorter time than the primary relay; this is detected in the system as a violation.

Case 2. 14-bus IEEE system with DG

The introduction of DG in the system brought variations of the I_{sc} at certain points; in the sensitivity analysis for this case, a total of 40 coordination pairs were identified. Table 4.3 shows the resulting times for primary and

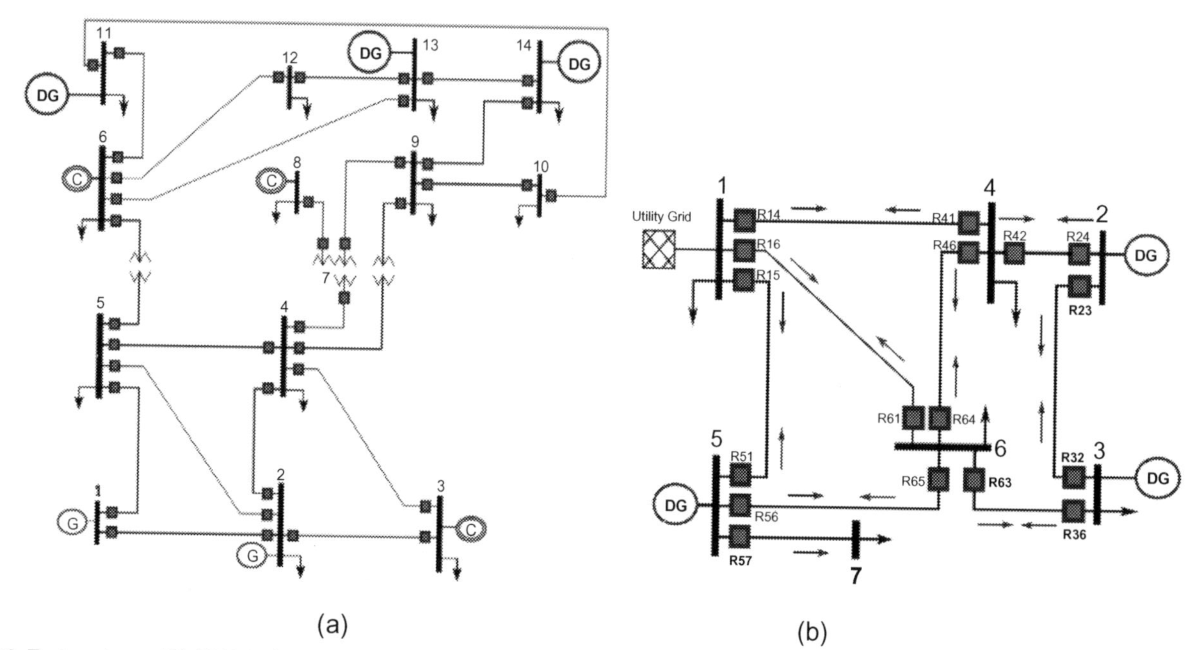

Fig. 4.11 Test systems. (A) IEEE-14 bus test system, (B) MG test system.

Table 4.2 Results obtained for the 14-bus IEEE system in normal operation.

Relay					Relay				
Primary	**Backup**	t_p	t_b	**CTI_c**	**Primary**	**Backup**	t_p	t_b	**CTI_c**
23	12	0.178	0.660	0.482	42	34	0.227	0.605	0.378
24	12	0.248	0.658	0.411	42	54	0.227	0.533	0.306
25	12	0.339	0.657	0.317	52	15	0.201	0.673	0.472
34	23	0.321	0.622	0.301	52	45	0.201	0.546	0.345
45	24	0.333	0.641	0.308	43	24	0.254	0.697	0.443
45	34	0.333	0.643	0.310	43	54	0.254	0.567	0.313
612	116	0.663	0.971	0.309	54	15	0.361	0.676	0.315
613	116	0.548	0.887	0.339	54	25	0.361	0.679	0.318
613	126	0.548	0.869	0.320	116	1011	0.439	0.296	−0.143
910	149	0.519	0.885	0.366	126	1312	0.156	0.491	0.335
914	109	0.172	0.605	0.433	136	1213	0.147	0.453	0.306
1011	910	0.226	0.629	0.403	136	1413	0.147	0.650	0.503
1213	612	0.439	0.745	0.306	109	1110	0.436	0.195	−0.241
1314	613	0.215	0.785	0.570	149	1314	0.498	0.270	−0.227
1314	1213	0.215	0.472	0.257	1110	611	0.114	0.657	0.543
21	52	1.122	1.444	0.323	1312	613	0.454	0.756	0.301
51	25	0.405	0.811	0.406	1312	1413	0.454	0.778	0.323
51	45	0.405	0.708	0.303	1413	914	0.461	0.196	−0.265
32	43	1.989	0.343	−1.645					
Number of Violations (NV)					5				
Differential Evolution Computation Time (s)					DE: 25.33				

backup relays obtained in the IEEE 14 node system with DG penetration applying the conventional coordination method. The presence of DG introduces a more complex scenario since the additional contribution may cause a loss of coordination. Table B in the Appendix section shows the I_{sc} and I_{Load} values for this analysis.

Case 3. MG connected

Table 4.4 shows the resulting times for primary operation and backup obtained in the MG system connected to the utility grid applying the conventional coordination method. There are no violations, but the operating times of the relays are increased by having similar short-circuit current values. Table C in the Appendix section shows the I_{sc} and I_{Load} values for this analysis.

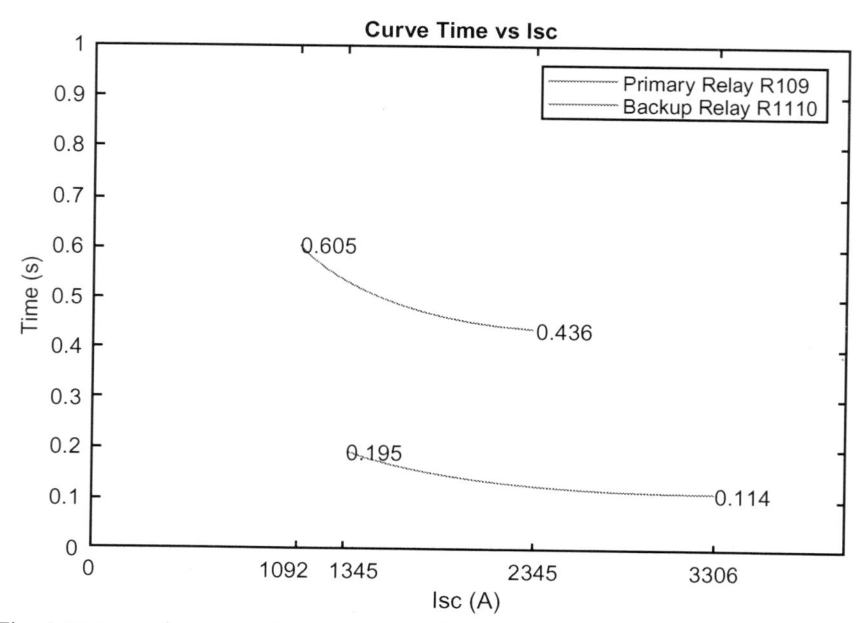

Fig. 4.12 Loss of coordination of one pair of relays.

Table 4.3 Results obtained for the IEEE system of 14 buses with DG penetration.

Relay					Relay				
Primary	**Backup**	t_p	t_b	**CTI_c**	**Primary**	**Backup**	t_p	t_b	**CTI_c**
23	12	0.284	0.775	0.491	51	45	0.137	0.833	0.696
24	12	0.381	0.773	0.393	32	43	0.348	0.651	0.303
25	12	0.441	0.771	0.331	42	34	0.412	0.955	0.544
34	23	0.501	0.999	0.498	42	54	0.412	0.759	0.347
45	24	0.480	0.960	0.480	52	15	0.386	0.912	0.526
45	34	0.480	1.002	0.522	52	45	0.386	0.697	0.312
611	126	0.212	1.494	1.281	43	24	0.494	1.057	0.563
611	136	0.212	0.517	0.305	43	54	0.494	0.798	0.304
612	116	0.134	0.647	0.513	54	15	0.528	0.912	0.384
612	136	0.134	0.463	0.329	54	25	0.528	0.828	0.300
613	116	0.301	0.632	0.331	116	1011	0.501	0.096	−0.405
613	126	0.301	0.602	0.301	126	1312	0.448	0.119	−0.329
910	149	0.529	0.855	0.326	136	1213	0.276	0.599	0.324
914	109	0.847	0.147	−0.701	136	1413	0.276	0.637	0.361
1011	910	0.088	0.605	0.517	109	1110	0.142	0.570	0.427
1213	612	0.586	0.148	−0.438	149	1314	0.720	0.166	−0.554
1314	613	0.140	0.444	0.305	1110	611	0.528	0.246	−0.282
1314	1213	0.140	0.652	0.513	1312	613	0.118	0.420	0.301
21	52	1.324	1.648	0.324	1312	1413	0.118	0.661	0.543
51	25	0.137	1.007	0.870	1413	914	0.521	0.957	0.435
Number of Violations (NV)					6				
Differential Evolution Computation Time (s)					DE: 28.12				

Table 4.4 Results obtained for the MG system connected to the utility grid.

Relay					Relay				
Primary	Backup	t_p	t_b	CTI_c	Primary	Backup	t_p	t_b	CTI_c
23	42	0.253	0.850	0.597	61	56	0.437	0.741	0.304
24	32	0.192	0.645	0.453	32	63	0.213	0.771	0.558
36	23	0.210	0.760	0.550	42	14	0.350	1.365	1.015
46	14	0.161	0.904	0.742	42	64	0.350	0.887	0.537
46	24	0.161	0.536	0.375	63	46	0.358	0.697	0.339
56	15	0.192	0.603	0.412	63	56	0.358	1.110	0.752
41	24	0.364	0.667	0.302	64	36	0.267	0.585	0.318
41	64	0.364	0.785	0.421	64	56	0.267	0.952	0.685
51	65	0.511	0.960	0.449	65	36	0.339	0.762	0.422
61	36	0.437	0.738	0.301	65	46	0.339	0.915	0.575
61	46	0.437	0.750	0.313					
Number of Violations (NV)					0				
Differential Evolution Computation Time (s)					DE: 12.39				

Case 4. MG islanding

Table 4.5 shows the resulting times for primary operation and backup obtained in the MG system operating in island mode applying the conventional coordination method. The nonconnection of the public network and the operation on the island brought variations of the I_{sc} at certain points, in the sensitivity analysis for this case a total of 17 coordination pairs resulted. As can be seen in Table 4.5, there are no violations; this is mainly due to the

Table 4.5 Results obtained for the MG system operating in island mode.

Relay					Relay				
Primary	Backup	t_p	t_b	CTI_c	Primary	Backup	t_p	t_b	CTI_c
16	51	3.921	4.495	0.573	61	56	0.965	2.549	1.583
23	42	0.328	0.913	0.585	32	63	0.371	2.418	2.047
24	32	0.268	1.244	0.976	42	64	0.358	1.155	0.797
36	23	0.339	1.062	0.723	63	46	1.049	1.610	0.561
46	24	0.240	0.837	0.597	63	56	1.049	4.004	2.955
41	24	0.567	1.009	0.442	64	36	0.274	1.054	0.781
41	64	0.567	1.072	0.506	64	56	0.274	3.022	2.748
51	65	0.704	1.368	0.664	65	36	0.356	1.268	0.911
61	36	0.965	1.274	0.309					
Number of Violations (NV)					0				
Differential Evolution Computation Time (s)					DE: 11.95				

small network dimensions, but very high operating times are presented as a negative aspect, which can be harmful for protection and voltage quality purposes. Table D in the Appendix section shows the I_{sc} and I_{Load} values for this analysis.

3. Enhanced DOCR coordination

The proposed coordination method consists of defining operating time intervals for the relay setting locations, these times must guarantee that the primary protection, backup, and sensitivity functions are obtained. Each relay must have a time curve that guarantees compliance with the restrictions in each interval, the use of nonstandardized curves allows the desired curve for each relay to be obtained. With this, a functional independence between relays is achieved, since the operating time does not depend on the time of another relay, as in Eq. (4.8), where the backup time t_b is a function of the primary time t_p. With this, a less algorithmically demanding formulation is obtained in the optimization problem since the coordination constraints for *CTI* are eliminated and only the operating time within each interval is minimized for each relay.

The increased tripping times of DOCRs are mostly before faults located far from their location; this problem can be particularly accentuated when standardized inverse-time curves are used or when only maximum faults are considered to carry out relay coordination. The use of nonstandardized inverse-time curves, the model and implementation of optimization algorithms capable of carrying out the coordination process, are proposed methodologies focused on the overcurrent relay performance improvement. These techniques may transform the typical overcurrent relay into a more sophisticated one without changing its fundamental principles and advantages. Consequently, a more secure and still economical alternative can be obtained, increasing its implementation area. The use of nonstandardized inverse-time curves would reduce the tripping time for lower fault currents and consequently improve the coordination process. Different levels of short-circuit current are computed to validate the robustness of the algorithm. This equation has five settings like degrees of freedom to give a solution to the coordination of each relay.

3.1 Interval coordination method

In the conventional coordination method, the functional dependence between relays is presented due to compliance with the *CTI*, where an increase in the operation time of a relay will affect its backups due to Eq. (4.8). Also, when the setting of a relay is modified, it is necessary to re-coordinate all relays. In the proposed method, time intervals are defined for each relay to satisfy its primary and backup protection functions. To obtain the sequence of operation between the primary and backup functions, the intervals are defined equally for all relays. In Fig. 4.13, for each fault current, the operation of the relays will be carried out complying with the defined time intervals, guaranteeing coordination. On the horizontal axis, the location of the time curve for compliance with the interval depends on the I_{pickup} and the I_{sc} of each relay. In dynamic operating conditions, the relays involved will have the same performance as conventional relays.

The problem is formulated as a minimization optimization function because it is necessary to determine the coefficients of the time curve that results in the shortest times for all intervals. To achieve this, time curves with greater flexibility are used to reduce coordination violations. These non-conventional curves are based on the general model having similar shapes to conventional curves, thus meeting the protection requirements. This formulation in coordination is less algorithmically demanding, since it is not

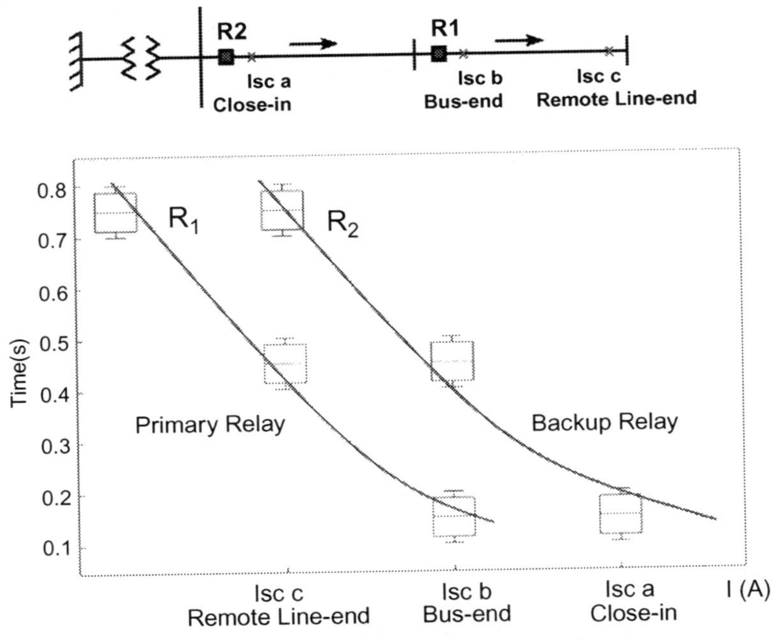

Fig. 4.13 Time intervals established for primary and backup functions.

necessary to find a solution that combines the operating time of all functionally dependent relays and their coordination constraints. In the proposed function, only the operating time within each interval is minimized for each relay. Then, only relays affected by load current or fault current changes will need to be updated. By using curves with higher degrees of freedom, the variables in the coordination problem are increased; however, this increase is not dominant in the results obtained in the test system analyzed.

The proposed method of interval coordination does not solve the crossing of curves since this is caused by the nature of the electrical network and the functional limitations of the overcurrent protection principle. In conventional coordination methods, the loss of coordination is not detectable when coordination is only performed at a single point of fault, such as coordination in radial networks.

In meshed networks, it is important that the coordination for faults is carried out both at the beginning and at the end of the line. There are violations in the results of the coordination, and it is often not possible to identify which relay initiates the crossing of curves, until the coordination can be graphed. In the interval method, the relay that does not comply with the intervals is presented directly in the violation and is easily detectable. In Fig. 4.14, the crossing of curves between a coordination pair is shown; for the conventional coordination method 14a, it can be detected because the crossing is close to the coordination current, but 14b will not be detected. In the proposed interval method, the R_1 relay does not meet its first interval in both cases. In these cases, the protection principle can be changed by differential relays or distance relays, replacing the problem relay.

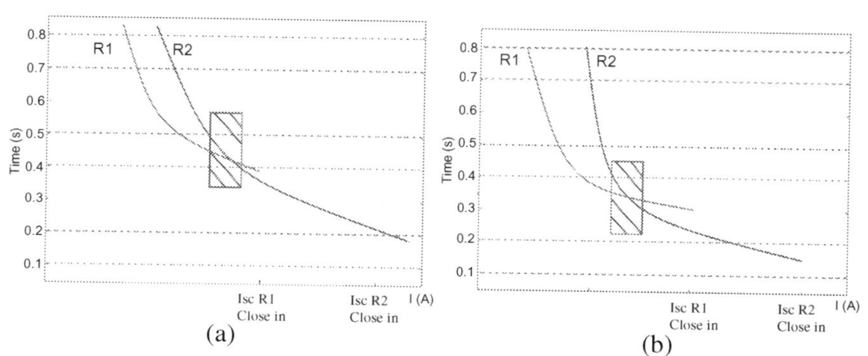

Fig. 4.14 Crossing of operation curves. (A) Crossing in the preset operating time interval and with the calculated I_{sc}. (B) Crossing outside the preset operating time interval and with another I_{sc}.

3.2 Optimization problem

The coordination problem is raised to minimize the operation time of each relay to comply with the three-time intervals (4.9).

$$OF = min\left\{\sum_{\mu=1}^{N}\left(T(I_{sc\ a})_{\mu} + T(I_{sc\ b})_{\mu} + T(I_{sc\ c})_{\mu}\right) + NV\right\} \tag{4.9}$$

where N is the number of relays, $T\ (I_{sc})$ are the relay operation times evaluated with nonconventional time curves. To improve the results. it is necessary to penalize the infractions to the inequality restrictions. the NV parameter is included, which is the number of violations that are presented in the search for the functions.

Restrictions. The objective function is achieved if the restrictions of the relay parameters and their limitations are satisfied. The time restrictions are established for each point of interest in the time intervals; the same curve must satisfy the three regions in a minimum time.

$$T(I_{Isc\ a})^{min} \leq T(I_{Isc\ a}) \leq T(I_{Isc\ a})^{max} \tag{4.10}$$

$$T(I_{Isc\ b})^{min} \leq T(I_{Isc\ b}) \leq T(I_{Isc\ b})^{max} \tag{4.11}$$

$$T(I_{Isc\ c})^{min} \leq T(I_{Isc\ c}) \leq T(I_{Isc\ c})^{max} \tag{4.12}$$

The time intervals are defined in compliance with the reliability and safety requirements. Depending on the protection scheme, the operating time requirements of DOCRs may change, mainly in their primary function. For example, for schemes where the DOCR backs up another relay, operating times can be increased by relaxing the requirement on the time curve. It is common for the internal policies of each electric company to define the allowable trip times; these values can be specified as restrictions on the time intervals defined in this proposal.

Equality restrictions of nonstandard time curves. In this chapter, we propose coordinating DOCRs considering five adjustable settings for three different levels of short-circuit current, obtaining nonstandardized inverse time curves. The proposed objective function considers close-in, bus-end, and remote line-end faults as frontier coordination currents. The use of nonstandard curves in DOCRs allows reduced operating times. The analytical expression of overcurrent time operation is show in Eq. (4.13):

$$\zeta(x) = \left[\frac{\mu_1}{\left(\frac{a}{b\cdot\mu_5}\right)^{\mu_3} - 1} + \mu_2\right]\cdot\mu_4 \tag{4.13}$$

where

$$\mu = (\mu_1, \mu_2, \mu_3, \mu_4, \mu_5)^T$$

Constraints:

$$a = \left(a^{min} : a^{max}\right)^T,$$
$$\mu_{1:5}^{min} \leq \mu_{1:5} \leq \mu_{1:5}^{max}, \quad \mu, a \in \mathbb{R}$$

The coefficients μ are considered as bounded variables in the Eq. (4.13) to obtain nonstandardized curves. It is common for μ_4 and μ_5 to be used as degrees of freedom to improve coordination (4.5); when the coefficients $\mu_{1:3}$ are also variables, nonstandardized curves are obtained (Carlos et al., 2015) (Table 4.6).

The ranges are defined for each coefficient of the curve for the algorithm to select the relay settings. The limits of variables μ are used to avoid time curves very different from those of overcurrent relays. These are based on the values used in the conventional curves (Table 4.1) to preserve the shape of the inverse time curves, as can be seen in Fig. 4.15.

For the bio-inspired optimization algorithms, the initial population is created by generating uniformly distributed random numbers; each number must be located within each setting boundaries. The structure of chromosomes is shown in Fig. 4.16, and it must be evaluated for the objective function.

3.3 Evaluation and presentation of the proposed interval coordination method

Table 4.7 shows the resulting times for primary operation and backup obtained in the 14-node IEEE system applying the proposed interval coordination method. In the simulations, the intervals used in Section 2.3 are

Table 4.6 Selection of parameters ranges.

	Boundaries	
Parameter	**Minimum**	**Maximum**
μ_4 (*TDS*)	0.1	3.0
μ_5 (*PSM*)	1.4	1.6
μ_1 (*A*)	0.01	28
μ_2 (*B*)	0	0.49
μ_3 (*p*)	0.01	2

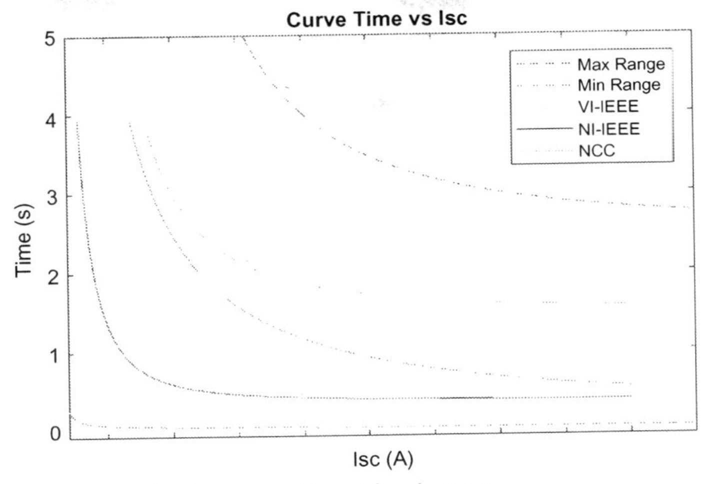

Fig. 4.15 Nonconventional curves within defined ranges.

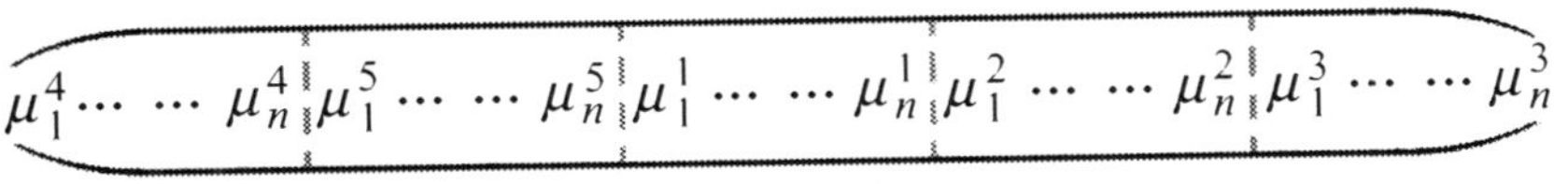

$$\mu_1^4 \cdots \cdots \mu_n^4 \,|\, \mu_1^5 \cdots \cdots \mu_n^5 \,|\, \mu_1^1 \cdots \cdots \mu_n^1 \,|\, \mu_1^2 \cdots \cdots \mu_n^2 \,|\, \mu_1^3 \cdots \cdots \mu_n^3$$

Fig. 4.16 Chromosome structure.

equal for caparison purposes; the operating times are analyzed showing that the reduction of *TDS* in conventional method not always minimize operating times. However, the minimal TDS is not always an advantage since more violations, or a loss of coordination can be presented. In addition, times can be obtained that are detrimental to the protection scheme. The results obtained with the proposed interval method shows that all relays are accomplished with the required time intervals defined: closing (0.1–0.2 s), end of bus (0.4–0.5 s), and remote end of line (0.7–0.8 s) to guarantee operation without violations and with lower operation times compared with results obtained in Case 1 (Table 4.2). Table A in the Appendix section shows the I_{sc} and I_{Load} values for this analysis. In Fig. 4.17, relay operating times and compliance with the preestablished operating interval are presented.

The comparison between coordination pairs of relays with conventional coordination and with the proposed coordination method is shown in Fig. 4.18. It can be observed that the conventional coordination pair presents longer operating times due to the functional dependence between relays. In

Table 4.7 Results obtained with proposed interval method for the 14-bus IEEE system in normal operation.

No.	Relay	Primary time (Close-in)	Backup time	
			Bus-end	Remote line-end
1	23	0.155	0.423	0.750
2	24	0.114	0.495	0.712
3	25	0.134	0.403	0.701
4	34	0.103	0.415	0.715
5	45	0.170	0.407	0.745
6	612	0.191	0.400	0.709
7	613	0.189	0.400	0.759
8	910	0.143	0.400	0.712
9	914	0.194	0.411	0.700
10	1011	0.184	0.401	0.701
11	1213	0.194	0.423	0.701
12	1314	0.196	0.443	0.714
13	52	0.161	0.402	0.729
14	43	0.153	0.463	0.781
15	54	0.192	0.408	0.713
16	116	0.138	0.439	0.701
17	126	0.103	0.403	0.729
18	109	0.176	0.400	0.701
19	149	0.191	0.448	0.704
20	1110	0.126	0.402	0.706
21	1312	0.125	0.401	0.704
22	1413	0.175	0.492	0.752
Number of Violations (NV)				0
Computation Time (s)				38.01

the case of the proposed method, each relay has its specific time curve to be able to comply with the specified time intervals. The results for the remaining cases and the evaluation in different operating conditions of the electrical network are shown in Section 5.

4. On-line coordination system

Before solving the coordination using an optimization method, it is of critical note to discuss the steps carried out to obtain the data used as input. The initial steps include the input of the data for the tested networks considering the conversion of these to a common system base. Typical data contain the number of buses, bus type, bus voltages, generators, shunt and line

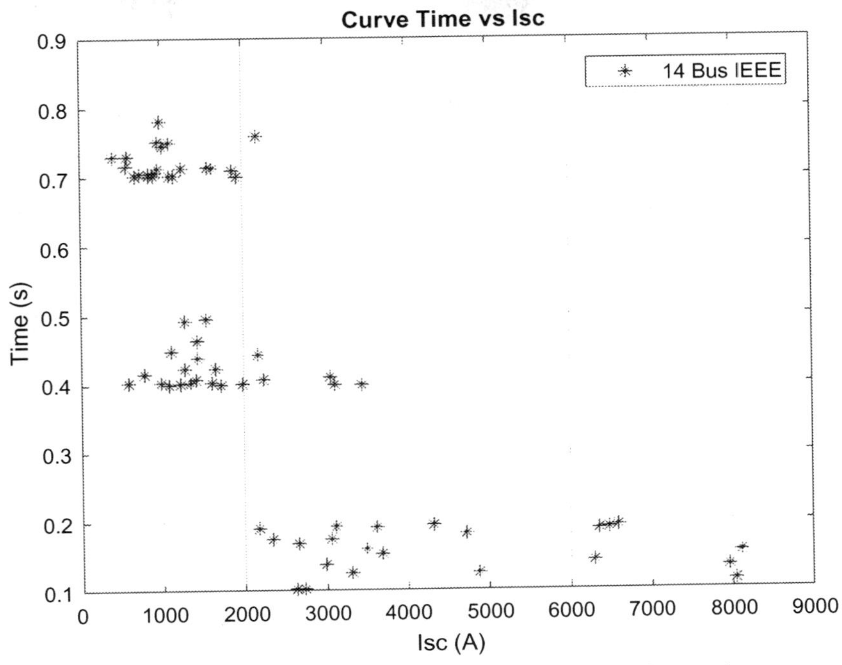

Fig. 4.17 Relay operating times for 14-bus IEEE system in normal operation.

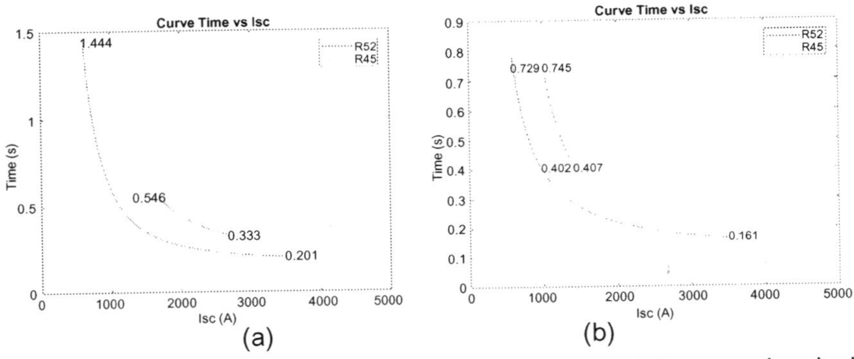

Fig. 4.18 Coordination relay times: (A) Conventional method and (B) proposed method.

impedances, generation, and load powers. The use of nodal matrices allows the topology of the electrical network in a steady state to be represented. The expressions are algebraic for the calculation of power flows and faults, the formulation is computationally efficient for topological and operational changes that must be considered during real-time operation. The output and input of elements such as generation sources, lines and loads must be considered; the intermittency of the RES can be considered the power injection in power flows and the reactance of the source for fault analysis.

4.1 Network topology

Obtaining nodal matrices is the basis for studies of the electrical network; thus, the nodal admittance matrix can be obtained by means of incidence matrices to obtain a computationally efficient method. In the same way, the nodal impedance matrix can be obtained by construction algorithms (Stagg & El-Abiad, 1968). These methods can consider mutual coupling scenarios and parallel lines very easily (Fig. 4.19).

The structure of the matrices allows updating the topological or operation variations by direct methods in the case of the Y_{nodal} and by inversion lemma in the case of the Z_{nodal}. The topological detection systems must have the position of the auxiliary contacts of the switch (52a) and the current measurement in that branch (Fig. 4.20). This information being verified at both ends will confirm the current topology and will allow the topological

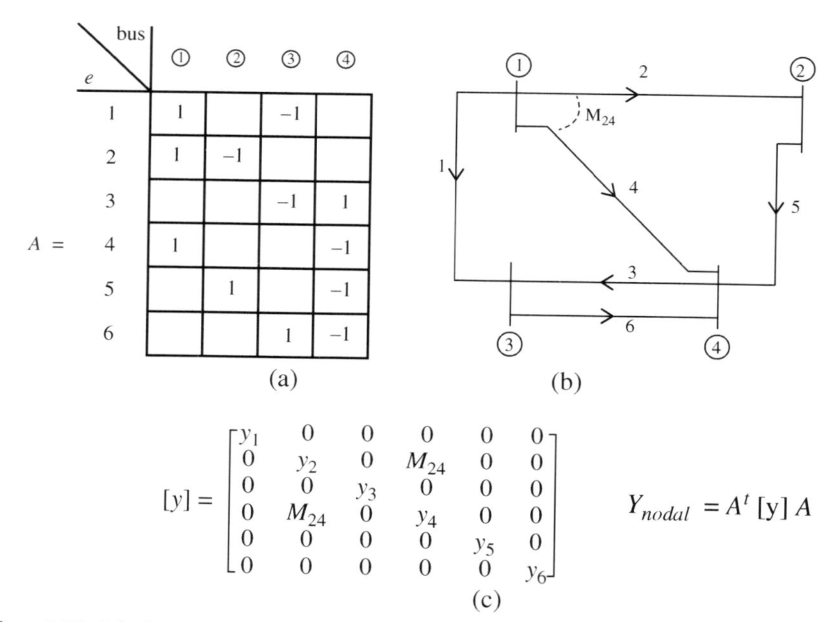

e \ bus	①	②	③	④
1	1		−1	
2	1	−1		
3			−1	1
4	1			−1
5		1		−1
6			1	−1

$$[y] = \begin{bmatrix} y_1 & 0 & 0 & 0 & 0 & 0 \\ 0 & y_2 & 0 & M_{24} & 0 & 0 \\ 0 & 0 & y_3 & 0 & 0 & 0 \\ 0 & M_{24} & 0 & y_4 & 0 & 0 \\ 0 & 0 & 0 & 0 & y_5 & 0 \\ 0 & 0 & 0 & 0 & 0 & y_6 \end{bmatrix} \qquad Y_{nodal} = A^t\,[y]\,A$$

Fig. 4.19 Admittance matrix. (A) Bus incidence matrix, (B) oriented network, (C) primitive admittance matrix.

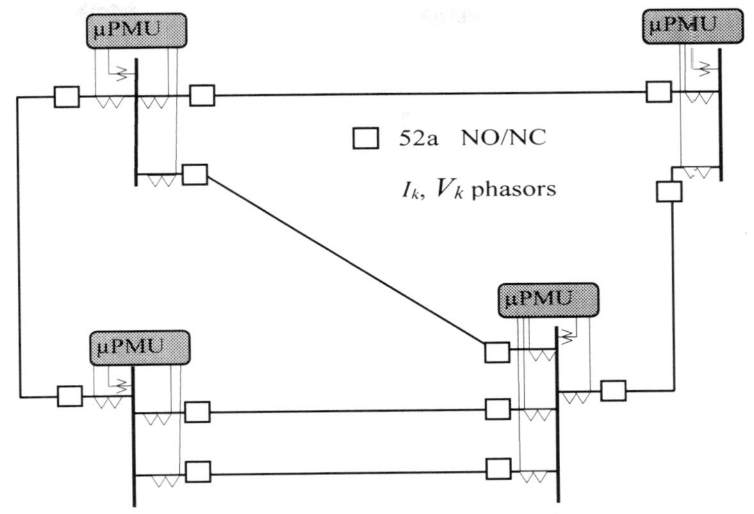

Fig. 4.20 Electrical network information for topology record.

modifications to be represented in nodal arrays. The voltage phasor and the sample time will allow to obtain the directionality of the measured current.

4.2 Load flow and fault current analysis

Load flow analysis is used to solve the nonlinear power flow equations to obtain node voltages and currents flowing in each direction of the network lines. The obtained voltages at each network bus and at each branch current are used to get the line loads each relay sees during the steady-state condition of the interconnected system. DOCR relays require the pickup currents to be set above the protected line maximum load current to prevent any unnecessary trips due to overload or transformer inrush currents To avoid false operations due to power transfer before the exit of some elements, it is necessary to carry out an analysis of contingencies $n-1$. In smart grids, the measurement obtained by µ-PMU is available, avoiding the execution of iterative flow algorithms; however, for contingency analysis it is necessary to run load flows algorithms. For identification purposes, all relays are assigned a name based on the network topology. If the relay is located between buses 1 and 4, with bus 1 being the closest to the given relay, the assigned relay name is *R14*.

The different current levels seen by each relay on interconnected systems increase the difficulty in conventional coordination methods, especially if there is more than one CPs formed by the same relay. On the other hand, the method proposed by intervals guarantees compliance with the times in each interval through the coefficients that determine the shape of the time curve. Routine fault analysis is used to determine the maximum and minimum fault currents for each location of interest. The maximum fault current is calculated assuming that a bolted three-phase fault (Zf=0) occurs at bus nearest to the relay, which is assumed to be of the same magnitude as a fault occurring directly in front of the relay. Minimum fault currents are used for a sensitivity analysis (4.3) and are obtained for the far end of the adjacent line. The faults are simulated using the Z_{th} of the diagonal of impedance nodal matrix. The far end line can be simulated like distribution systems, adding the impedance line to Z_{th} because the remote end of the line is considered open.

4.3 Special considerations

The operation of DOCRs is very simple, which means that this protection scheme can be widely used, especially on interconnected systems. The operation of the relay is triggered when the measured current cause a direction control unit and an overcurrent unit to operate after its magnitude surpasses its assigned pickup current. However, special considerations must be made in the coordination formulation; equipment protection like transformer, reactors, or capacitor bank can be included in the formulation. Also, parallel lines must be included because the formation of nodal matrices results in the equivalent parallel line, losing the information from the relays.

The coordination uses each relay in the network, commonly excluding those that protect power transformers because the use of differential protection schemes. On the other hand, the backup function of transformers and shunt elements, such as capacitors and reactors, can be included in the formulation as additional restrictions. If the branch is a transformer or a double line circuit, this must be declared in the input data. The algorithm will activate the procedure for setting relays for transformer protection to back up differential protection or primary protection.

4.3.1 Setting of relays for transformer protection

The application of DOCRs as transformer protection can be located depending on the capacity (IEEE C37.91-2000). The limit considered is 10MVA, when the capacity of the transformer is lower, only the overcurrent protection is used, when it is higher, the differential protection is added and the backup protection is the overcurrent protection.

Two winding power transformers < 10 MVA. Protection for transformers less than 10 MVA is basically overcurrent, these schemes must ensure the relay operation for internal transformer faults and guarantee backup for faults in the feeders (Fig. 4.21).

Protection on primary side, 51H. Primary transformer protection must operate for transformer faults, have coordination with low-side protection equipment and avoid operation under hot load pickup, like inrush current on instantaneous reclosing of source-side circuit breakers and cold load pickups. The setting of the pickup current of the primary protection must be:

For Fuse: 150% and 175% of low-speed *E* type high voltage side primary current

For Relay: 125%–150% of maximum kVA nameplate rating of a transformer is common

Both protections must coordinate with the damage curve of the transformer and allow the magnetization current, being greater between 8 and 12 times the nominal current in a time greater than 0.1 s. Only the application of relays for the generation of restriction functions is considered.

$$I^{ij}_{pickup\ 51H} = I_{load\ rating} * PSM_{Trans} \tag{4.14}$$

$$1.25 \leq PSM_{Trans} \leq 1.5 \tag{4.15}$$

$$T^{ij}_{relay} \geq 0.1 \text{ for } I_{inrush} = 8 * I_{nom} \tag{4.16}$$

The relay must protect the transformer for internal faults; to achieve this, its time curve must be less than the damage curve and greater than the inrush point. To guarantee the protection function of the transformer, it is only necessary to verify points *1*, *2*, and *4* for Cat II–IV and points *1* and *4* for Cat I (Fig. 4.22). This verification is carried out by determining the times of the damage curves for the transformer and its category as shown in Table 4.8.

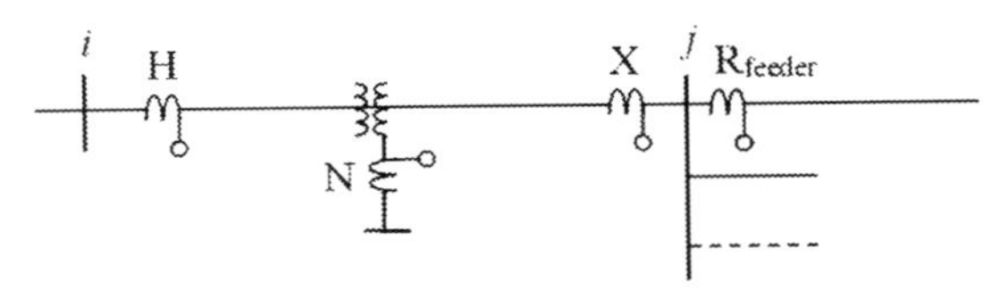

Fig. 4.21 Protection scheme for power transformer less of 10 MVA.

Table 4.8 ANSI curve points.

Point	Category of the transformer	Time (s)	Current (A)
1	I	$t = 1250 * (Z_t)^2$	$I_f = \frac{In}{Z_t}$
	II	$t = 2$	$I_f = \frac{In}{Z_t}$
	III, IV	$t = 2$	$I_f = \frac{In}{Z_t + Z_s}$
2	II	$t = 4.08$	$I_f = 0.7 * \frac{In}{Z_t}$
	III, IV	$t = 8$	$I_f = 0.5 * \frac{In}{Z_t + Z_s}$
3	II	$t = 2251 * (Z_t)^2$	$I_f = 0.7 * \frac{In}{Z_t}$
	III, IV	$t = 5000 * (Z_t + Z_s)^2$	$I_f = 0.5 * \frac{In}{Z_t + Z_s}$
4	I, II, III, IV	$t = 50$	$I = 5 * In$

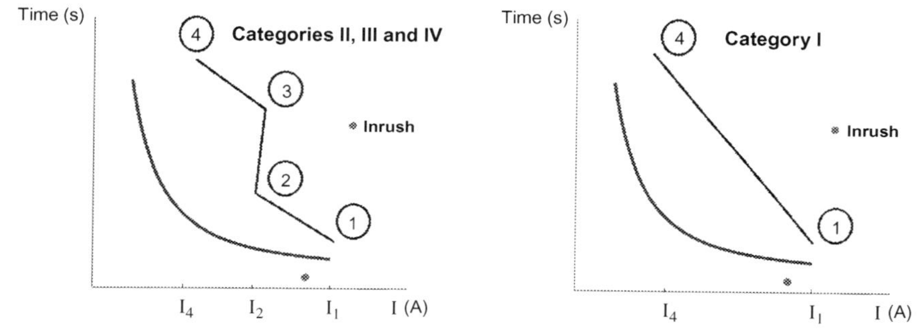

Fig. 4.22 Transformer damage curve category.

For Cat II, III, and IV:

$$T^{ij}_{relay}(I1) > T1 \text{ and } T^{ij}_{relay}(I2) > T2 \text{ and } T^{ij}_{relay}(I4) > T4 \tag{4.17}$$

For Cat I:

$$T^{ij}_{relay}(I1) > T1 \text{ and } T^{ij}_{relay}(I4) > T4 \tag{4.18}$$

Phase overcurrent protection on secondary side, 51X. The relays must coordinate with the relays installed in the R_{Feeder} feeders and must operate with time delay for three-phase or two-phase faults in the feeders ($I^{3\varnothing}_{sc\ x}$), maintaining the CTI of the slowest protection of the feeders.

$$I^{ij}_{pickup\ 51X} = I_{load\ rating} * 2.0 \tag{4.19}$$

$$0.5 \leq T^{ij}_{relay\ 51X}\left(I^{3\varnothing}_{sc\ x}\right) \leq 0.8 \tag{4.20}$$

Two winding power transformers ≥10MVA. The primary protection function of transformer is differential protection. Adjustment considerations do not affect the coordination of DOCRs and are not described. Transformer backup and external fault backup functions are performed by DOCRs.

Phase overcurrent protection on primary side, 50/51H. The function of the instantaneous element (50H) is to protect the transformer without overreaching the feeders on the secondary side (Fig. 4.21), two conditions must be met:

$$I_{load\ rating}{*}10 < I^{ij}_{pickup\ 50H} \text{ and } 2{*}I^{3\varnothing}_{sc\ x} < I^{ij}_{pickup\ 50H} \tag{4.21}$$

Protection on primary side, 51H. It must allow the primary protections of the bank and the feeders to operate first. It must operate with time delay *CTI* for faults on the low side of the bank or on feeders and must operate as transformer overload protection.

$$I^{ij}_{pickup\ 51X} = I_{load\ rating}{*}PSM_{51H} \tag{4.22}$$

$$2.0 \leq PSM_{51H} \leq 2.2 \tag{4.23}$$

The time curve must be below the damage curve (for frequent fault) of the transformer. The operating time must be between 0.8 and 1 s for a three-phase fault on the low voltage bus.

$$0.8 \leq T^{ij}_{relay\ 51H}\left(I^{3\varnothing}_{sc\ x}\right) \leq 1.0 \tag{4.24}$$

Transformer secondary side backup protection, 51X. It must allow the bank's primary protections and feeder protections to operate first. It must operate with a time delay for three-phase or two-phase faults in the feeders (Fig. 4.21).

$$I^{ij}_{pickup\ 51X} = I_{load\ rating}{*}2.0 \tag{4.25}$$

$$0.5 \leq T^{ij}_{relay\ 51x}\left(I^{3\varnothing}_{sc\ x}\right) \leq 0.9 \tag{4.26}$$

4.3.2 Parallel lines

In the case of parallel lines (Fig. 4.23), the corresponding pair of relays will be assigned to the equivalent branch. The adjustment will be made considering the fault current and load current in each line in proportion to the impedance ratio of each line; the most common case is that the lines have the same

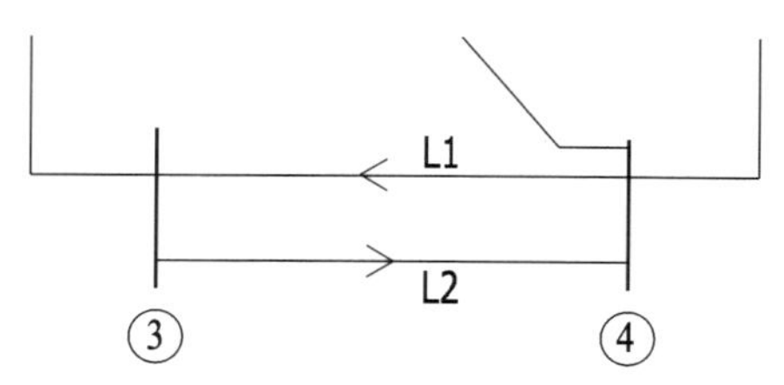

Fig. 4.23 Parallel lines.

impedance. With these values, the pickup current in each relay is determined and the corresponding coordination pairs are generated.

$$I_{pickup}^{L1} = I_{load\ L1} * PSM \tag{4.27}$$

$$I_{pickup}^{L2} = I_{load\ L2} * PSM \tag{4.28}$$

Coordination between relays *R43* and *R34* on different lines will be established by coordinating both relays (Fig. 4.23) using the conventional coordination method:

$$T_{R43}^{L1} = T_{R34}^{L2} + CTI \text{ for Fault } L34 \text{ end line open} \tag{4.29}$$

$$T_{R34}^{L1} = T_{R43}^{L2} + CTI \text{ for Fault } L43 \text{ end line open} \tag{4.30}$$

With the proposed method, only compliance with the intervals will be necessary.

The fault analysis for relay setting is done by considering the line-end open; the fault contribution is from only one end. This criterion is applied under the consideration that the probability that relays located at each end failing to operate is very low. In the case of faults on parallel lines, the contribution from the remote end is given by the nonfault line. For this reason, the fault calculation for the coordination of relays must be carried out considering the remote end connected with the impedance of a single line, the Y_{nodal} must consider this increase in the impedance of the branch or use the inversion lemma for modify the Z_{nodal} (Fig. 4.24).

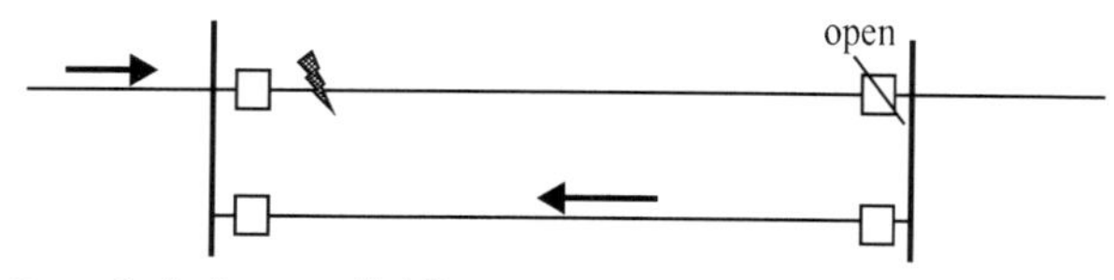

Fig. 4.24 Fault analysis for parallel lines.

4.3.3 Radial lines detection

Radial line protection is located at the source end; therefore, this condition must be recognized by the algorithm and reflected in the formulation. The connection of a single branch to a node is an indicator of it being the last node of the load end. The Y_{nodal} construction algorithm must indicate when a single branch is connected to a node on the diagonal of the matrix (4.31), or when there is only one mutual element (y_{45}) in the column vector or row vector, Fig. 4.25. From this node, the algorithm performs a search toward the next value on the diagonal or in the next column vector that has the previous branch and an additional branch (y_{34}). This process detects lines with a radial nature, ending when there is an element of the diagonal with three or more connected branches. The algorithm's relay assignment will be undertaken from the source node, omitting the remote end relay.

$$Y_{nodal} = \begin{bmatrix} (y_{13}+y_{12}) & y_{12} & y_{13} & 0 & 0 \\ y_{12} & (y_{23}+y_{12}) & y_{23} & 0 & 0 \\ y_{13} & y_{23} & (y_{13}+y_{23}+y_{34}) & -y_{34} & 0 \\ 0 & 0 & -y_{34} & (y_{34}+y_{45}) & -y_{45} \\ 0 & 0 & 0 & -y_{45} & (y_{45}) \end{bmatrix} \tag{4.31}$$

4.4 Sensitivity analysis

The sensitivity analysis is performed to remove all of the coordination pairs that, due to system restrictions, will not be able to coordinate. Backup relays should have the ability to detect minimum fault currents that occur at the far end of its protected line. Eq. (4.32) is used to analyze each coordination pair to determine whether coordination can be achieved. The coordination pairs that do not overcome the sensitivity analysis are relays with either high or

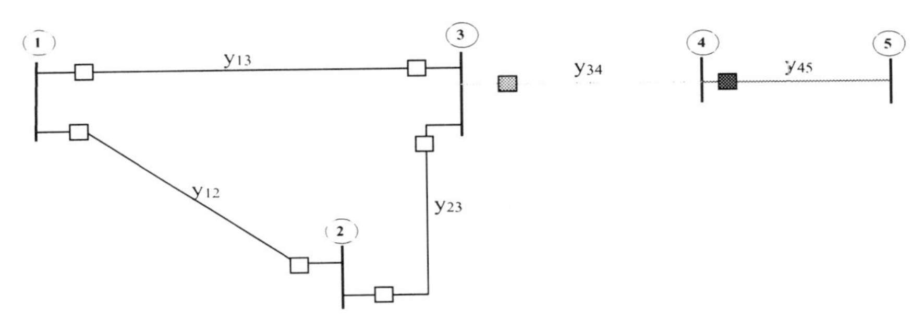

Fig. 4.25 Radial line detection example.

unacceptable operating times. The lack of sensitivity is a limitation of the overcurrent principle used; the use of distance relays for backup functions can be an alternative. In cases where there are numerous relays that do not pass the filter (4.32), a lower percentage of coverage of the backup function can be considered only to obtain a relay setting (Fig. 4.26). For example, 80%; this percentage will depend on the impedance of the backup line. This scheme should be supplemented with transferred trip schemes to ensure that 100% of the line is backed up.

$$S_b^{NF} \geq \frac{V_{nom} / \left(Z_{line}^{backup} * 0.8 + Z_{th} \right)}{I_{pickup}} \quad (4.32)$$

The pseudocode of Fig. 4.27 is used for evaluating the sensitivity and active o disable relays. This step is convenient for coordination since, when a relay lacks sensitivity, it will be an algorithmic violation that cannot be avoided, thus increasing the execution time of the optimization algorithm.

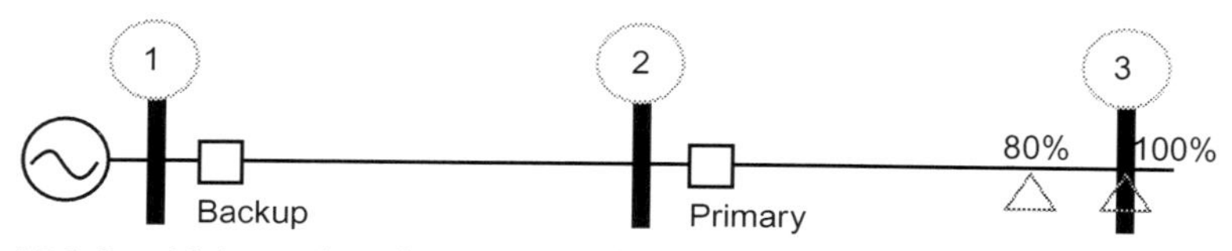

Fig. 4.26 Sensitivity analysis for minimum fault currents that occur at the far end of its protected line.

```
Step 1. Run initial routines
        Topology detection for nodal matrix construction
        Calculation of Fault current and power flow
Step 2. Three-phase and two-phase current seen by each relay
        Three-Phase Current through Relay – Irel_3P
        Two-phasic Current through Relay – Irel_2P
Step 3. Primary and Backup Relays Line Currents
        Primary Relays r= Start Node; s= End node;
        Backup Relays m= Start Node; n= End node;
        Current Line Primary Relay ILp(r,s)=abs(-Ybus(r,s)*(Vbus(r)-Vbus(s)))*IBase(r)
        Current Line Backup Relay ILb(m,n)=abs(-Ybus(m,n)*(Vbus(m)-Vbus(n)))*IBase(m)
Step 4. Sensitivity Calculation
        Sb=Irel_2P/(1.5*ILb(m,n))
Step 5. Determine which relay pairs can be coordinated
        Sensitivity Condition Scond=find (Sb > 1.5)
Step 6. Reduced Coordination Matrix
        Relays that they comply with the Sensitivity Irel_3Pc=Irel_3P (Scond)
        Relays that they comply with the Sensitivity Irel_2Pc=Irel_2P(Scond)
Step 7. Coordination Pairs (CPs) after Sensitivity analysis
        CPsn=CPs(Scond)
End
```

Fig. 4.27 Pseudocode of sensitivity.

4.5 Formation of optimization constraints

The coordination of DOCRs in large interconnected systems is a highly constrained optimization problem that has been solved using various techniques. The restrictions limit the decision variables on its cost function making it even more complex to solve using conventional methods. The coordination viewed as an optimization problem requires the satisfaction of both hard and soft constraints on the formulation of its population and the objective function. The hard constraints are given for ranges on which its decision variables should operate while the soft constraints aid with assisting in the final decision to choose the best individual of a population.

Equality constraints. They are the expressions of the time curves of each relay shown in Table 4.6. The degrees of freedom of the curve will allow you to adjust to the coordination requirement by being incorporated into the on-line coordination algorithm; the parameters of the curve will be the output results.

In Fig. 4.28, the pseudocode for the equality restrictions is shown, the parameters that shape the nonconventional time curves are evaluated, as well as the formulation for the calculation of the operation times. There will be an equality equation for N active relays ($\zeta(x)^N$); each variable parameter of Eq. (4.13) will be bounded, generating corresponding inequality equations.

```
Step 1. Run initial routines
        Calculation of Fault current and power flow
Step 2. Adjustable settings for the Algorithm
        Relay Adjustable Settings AS=5
        Agents or individuals (Pack Size) PS=80
        Total Iteration Niter=1000
        Total times code repeated Crpt=1
        Proposed CTI=0.3
        Dimension of Decision Variables DIM = Nrelays*NS
Step 3. Upper and Lower Settings for Adjustable settings
        dial [min max]        d_lim= [0.1      3]
        k factor [min max]    fk_lim=[1.4    1.6]
        A [min max]           A_lim= [1       28]
        B [min max]           B_lim= [0.1   0.49]
        p [min max]           p_lim= [0.01     2]
Step 4. Primary and Backup Relays Information
        Primary Relays    Prdata=[NRelP   IscP   IloadP]
        Backup Relays     Brdata=[NRelB   IscB   IloadP]
Step 5. Primary Relay operating time
        tp= (A/ (((Prdata (2)/ (fk*Prdata (3))) ^p)-1) +B) *dial
Step 6. Backup Relays operating time
        tb= (A/ (((Brdata (2) / (fk*Brdata (3))) ^p)-1) +B) *dial
End
```

Fig. 4.28 Pseudocode of equality for time curve and I_{pickup} computation.

This could represent a computational increase but is offset by the increase in the feasibility of finding the solution. The correct operation of the relay is ensured by normally setting its pickup current greater than the maximum possible load current (I_{load}) and below the minimum fault current that can be seen on the protected line by using a plug setting multiplier (PSM) as shown in Eq. (4.33).

$$\left[I_{pickup} = I_{load}*PSM\right]^{N} \tag{4.33}$$

Inequality constraints. The *PSM* or μ_5 range should ensure that the relay discriminates between normal and abnormal conditions. The *TDS* or μ_4 is the multiplier setting used to increase or decrease the relay operation time; the range must be selected for obtaining the operation time relay adequate for protection proposes. The upper range is defined as 3 or 2 commonly; the lower range has had different values, ranging from 0.5 in the first jobs, to 0.1 and even 0.01. It is important to note that very small dials may result in very short operating times that can affect safety, triggering maneuvering operations or transients in the electrical network. The coefficients μ_{1-3} must be limited within ranges of the curve standards to maintain the shape of the time inverse overcurrent relays. The time constraints for each interval result in three inequality equations for each relay. This represents an algorithmic advantage since the evaluation of the restriction is for each relay (4.10–4.12).

In Fig. 4.29, the pseudocode for the inequality restrictions is shown. For comparison proposes, conventional coordination methods result in ϑ inequality restrictions for each CP. For example, the 14 IEEE test system has 50 pairs and 40 relays, and the 30 IEEE test system has 120 pairs and 82 relays.

$$\left(t_b - t_p \geq CTI\right)_{\vartheta} \tag{4.34}$$

The dependence of the times between each coordination pair results in a dependence between all relays. Modifying the setting of one relay will result in a domino effect affecting all relays. Instead, this is not the case for the interval method, because each relay has independent restrictions from other relays.

4.6 Optimization algorithms

The Differential Evolution algorithm operates through similar computational steps as employed by a standard evolutionary algorithm (EA). However, unlike traditional evolutionary algorithms, the Differential Evolution

```
Step 1. Run initial routines
Step 2. Adjustable settings for the Algorithm
        lim = [d_lim (2) fk_lim (2) A_lim (2) B_lim (2) p_lim (2);
               d_lim (1) fk_lim (1) A_lim (1) B_lim (1) p_lim (1)]
Step 3. Matrix with lower and upper Boundaries for the decision variables
        lub = repmat(lim,Nrelays)
Step 4. Upper and Lower Boundary Vectors
        lb=lub (2)   ub=lub (1)
Step 5. Initial Position
        Position =rand (PS, Number relays*AS);
Step 6. Aleatory dial, fk, A, B, P for initial generation
        dial=d_lim (1,1) +(d_lim (1,2)-d_lim (1,1)) *Position
        fk=fk_lim (1,1) +(fk_lim (1,2)-fk_lim (1,1)) * Position
        A=A_lim (1,1) +(A_lim (1,2)-A_lim (1,1)) * Position
        B=B_lim (1,1) +(B_lim (1,2)-B_lim (1,1)) * Position
        p=p_lim (1,1) +(p_lim (1,2)-p_lim (1,1)) * Position
Step 7. Coordination time Interval calculation
        CTI Calculation      CTI_c=tb - tp
        Error calculation    CTI_error=CTI_c - CTI
Step 8. Penalization for CTI errors
        Zero Penalization    CTI_e >= 0 & CTI_e <= 0.3
        Slightly Penalized   CTI_e > 0.3
        Highly Penalized     CTI_e < 0
Step 9. Number of Violations
        NV=NV + 1
End
```

Fig. 4.29 Pseudocode of inequality.

algorithm perturbs the current generation population members with the scaled differences of randomly selected and distinct population members. Therefore, no separate probability distribution must be used to generate the offspring. This characteristic means that the algorithm has fewer mathematical operations, and hence a shorter execution time than other algorithms.

In the Differential Evolution community, the individual trial solutions (which constitute the population) are called parameter vectors or genomes. Each parameter vector contains possible solution information. Crossover is a recombination (or reproduction). Sons or new individuals are formed by the recombination of the genes of certain parameter vectors called parents or target vectors. The newly formed individuals (sons) are called trial vectors. Mutation is a sudden change or perturbation with a random element. A mutant or donor vector is formed for each target vector. Selection is the process that keeps the population size constant over the subsequent generations by determining whether the target or trial vector survives to the next generation. Search space is the space of all feasible solutions. Each point in the search space represents a feasible solution and each feasible solution can be marked by its fitness value for the problem. This feasible solution is then a maximum or a minimum.

The algorithm starts with many sets of solutions; together, all of the parameter vectors form a population. Solutions are taken from a population and used to form a new one. By employing the selection process, you guarantee that the population will improve (relative to the minimization of the objective function) or remain the same in fitness, but never deteriorate. In Fig. 4.30, the pseudocode used for the implementation of this algorithm is presented.

4.7 On-line implementation diagram

The optimization algorithm that will be selected for on-line execution must show the robustness necessary for protection systems. To carry out the proposed coordination and the correct operation of the protections in the pre-established times, the adjustment update system must be implemented in a centralized mode with continuous real-time measurement processing and monitoring of the system topology. The time update before any topological change must be almost in real time, and the update by variation of the demand can be carried out in intervals of no more than 15 min, similar to the times used for the calculation of the demand. The latency of communication available in the electricity grid must be considered. The I_{pickup} is updated with the measured load current within the time interval considered, and the I_{sc} will be modified in case of a topological change or the connection of generation sources.

Step 1. Run initial routines
Step 2. Initialize parameters
Initial population and setting (NP – F – CR – Strategy)
Step 3. Coordination constraints
Degrees of freedom for dial and pickup
Degrees of freedom for non-conventional curves (A, B and p)
Time intervals for all coordination pairs of relays
Step 4. DE Algorithm

- ***Start***
- ***Randomly initialize the positions of all individuals***
- ***Evaluate the fitness function values of all individuals (Xi)***
- ***Mutation – Generate mutant vectors Vi, g+1***
- ***Crossover –Generate trial vector U i, g+1***
- ***Selection – Update individuals' position***
- ***Evaluation – Evaluate the aptitude of the new population***
- ***Run the algorithm again with the new population***

The algorithm ends if the convergence criterion is met
End

Fig. 4.30 Pseudocode Differential Evolution (DE).

In conventional coordination, the topological change of the power grid or the change of the load/fault current forces all coordination to be updated due to the functional dependence between relays. This adjustment update includes all relays, even though it is possible that many of them do not appreciate any variation in magnitudes. On the other hand, in the proposed method, the adjustment update will be only in the relays involved. Those who do not have an appreciable change in the currents will not have to be adjusted, reducing computing and communication times. In Fig. 4.31, the implementation of the proposed method is shown as a functional diagram. The update of the relay settings must be carried out in a stable state, so no change or update is

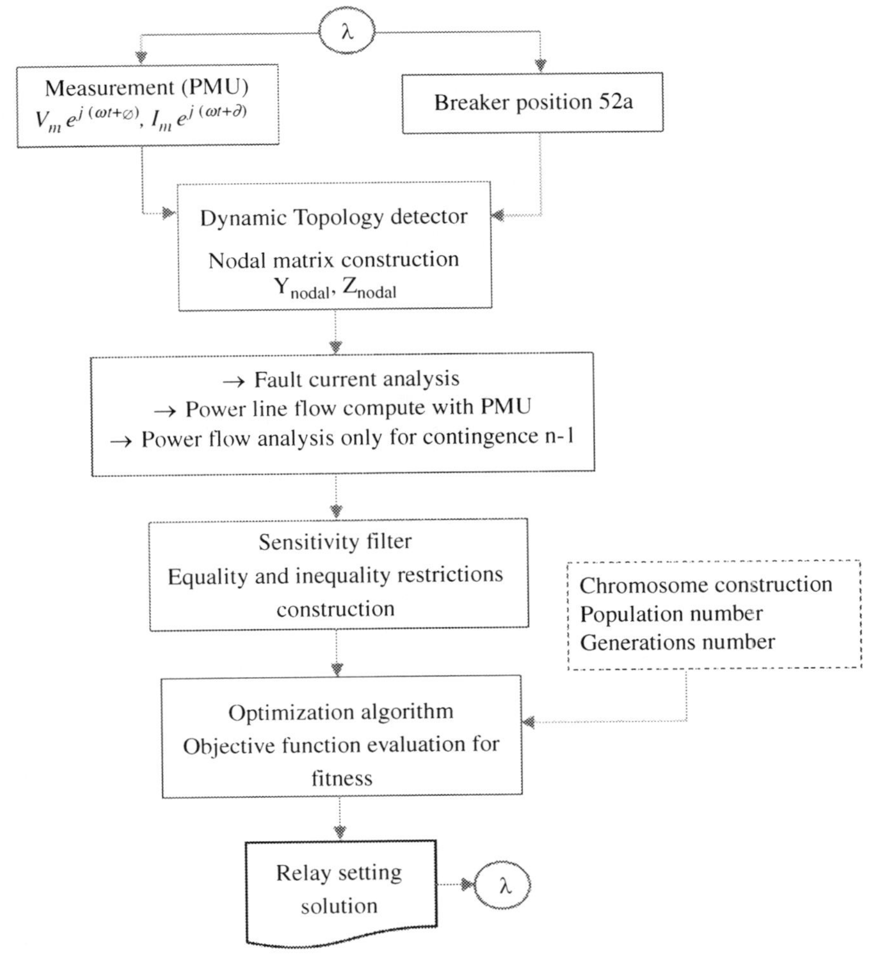

Fig. 4.31 Implementation of the proposed method of coordination through DE.

recommended during a fault condition to maintain the safety of the protection scheme. The use of units of measurement, for example μPMU, is recommended and can greatly reduce the amount of calculation. A sensitivity analysis is performed to eliminate relays that do not have the ability to detect minimal faults that occur at the remote line-end. The sensitivity filter to overcome for each relay is analyzed using Eq. (4.3).

5. Improve the resilience of the protection scheme in electrical systems

A more resilient protection scheme has a greater capacity to withstand disruptive events, reduce impacts, and accelerate recovery. In this work, several factors that integrate and complement the protection system for current energy networks are addressed. The results and simulations with the proposed time interval coordination method are presented below, as well as the analysis of various factors that influence the performance of the protection scheme.

5.1 IEEE 14 bus with DG penetration applying the proposed interval coordination method

At this point, the evaluation of the interval coordination method in the IEEE 14 bus system is carried out, for this case the penetration of DG sources is taken into account, as well as their contribution to the I_{sc} seen by the relays. The comparison is with Case 2 of Section 2.3, where conventional coordination was used. In Table 4.9, the results are displayed; it is clear how the proposed method guarantees the operation of the relays within the established time interval.

The influence of the DG and its contribution to the fault current do not prevent the algorithm for finding the necessary adjustments to guarantee good operating times. Fig. 4.32 shows the graphical time results.

5.2 Evaluation of the proposed method with the MG operating connected to the utility

Evaluation of the MG in association with the utility presented in case 3 of Section 2.3 is achieved using the proposed interval method. Table 4.10 shows the results obtained; it can be seen that the program obtains a time curve that complies with all intervals in all relays. If the backup times t_p in Table 4.4 and the times of the relays in the Bus end are compared, a significant reduction in operating times results. In Fig. 4.33, the relay times in each interval are shown.

Table 4.9 Results obtained with proposed interval method for the 14-bus IEEE system with DG penetration.

No.	Relay	Primary time (Close-in)	Backup time Bus-end	Backup time Remote line-end	No.	Relay	Primary time (Close-in)	Backup time Bus-end	Backup time Remote line-end
1	23	0.113	0.402	0.700	13	1314	0.170	0.401	0.777
2	24	0.102	0.485	0.700	14	52	0.150	0.476	0.714
3	25	0.104	0.491	0.754	15	43	0.167	0.447	0.727
4	34	0.141	0.440	0.718	16	54	0.186	0.400	0.738
5	45	0.170	0.400	0.714	17	116	0.188	0.400	0.776
6	611	0.186	0.401	0.736	18	126	0.130	0.420	0.785
7	612	0.186	0.401	0.746	19	52	0.150	0.476	0.714
8	613	0.187	0.400	0.719	20	109	0.171	0.403	0.731
9	910	0.161	0.406	0.779	21	149	0.185	0.413	0.75
10	914	0.186	0.420	0.726	22	1110	0.182	0.456	0.712
11	1011	0.162	0.401	0.785	23	1312	0.128	0.442	0.715
12	1213	0.168	0.404	0.700	24	1413	0.193	0.403	0.794
Number of Violations (NV)					0				
Computation Time (s)					39.57				

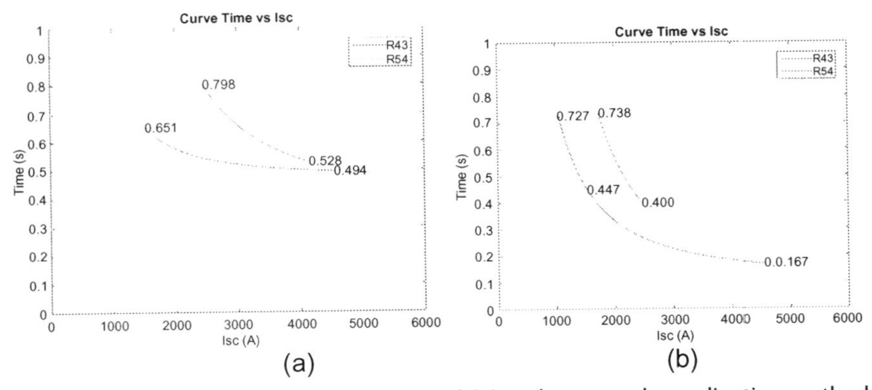

Fig. 4.32 Comparison between conventional (A) and proposed coordination method (B) with DG connection in IEEE 14 bus.

5.3 Evaluation of the proposed method with the MG operating in island mode

The islanding condition of MG results in emergency conditions or normal operating conditions, with high dynamic conditions for RES and for loads. The coordination performance is harshly evaluated because the relay setting

Table 4.10 Results obtained with proposed interval method for the microgrid connected to the system.

No.	Relay	Primary time (Close-in)	Backup time	
			Bus-end	Remote line-end
1	23	0.183	0.407	0.782
2	24	0.162	0.424	0.728
3	36	0.186	0.423	0.770
4	46	0.102	0.466	0.702
5	56	0.169	0.406	0.748
6	32	0.184	0.401	0.778
7	42	0.165	0.401	0.705
8	63	0.178	0.402	0.705
9	64	0.122	0.403	0.794
10	65	0.197	0.401	0.730
Number of Violations (NV)				0
Computation Time (s)				19.21

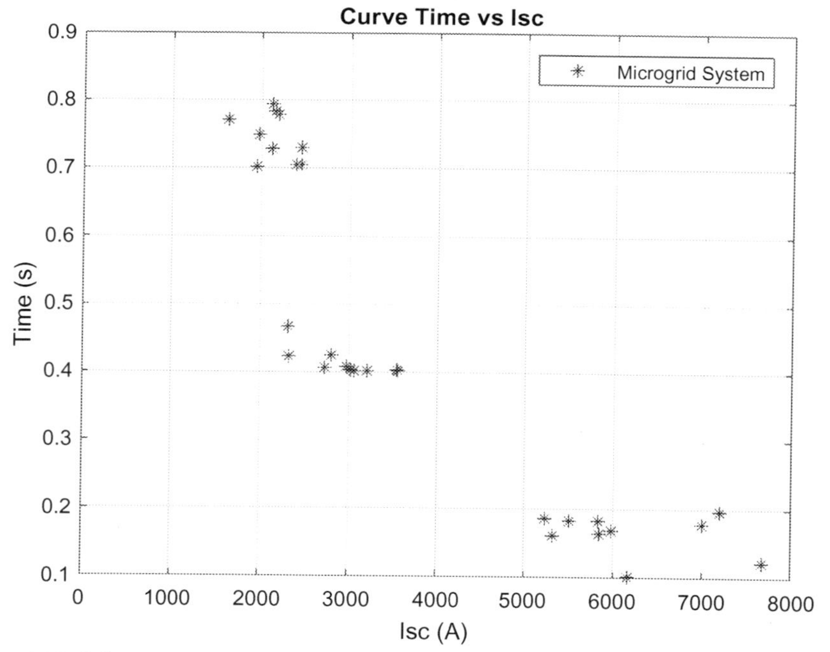

Fig. 4.33 Relay operating times for MG system connected to the system.

is subject to various dynamic conditions. Sensitivity is mainly compromised by low fault current values, expecting a high number of relays that do not meet the sensitivity criteria.

In Section 2.3, this condition was evaluated with conventional methods. The results were presented in Table 4.5. Although no violations were reported, the relay operating times were higher than for the interval proposed method presented in Table 4.11. In Fig. 4.34, the difference in the operation times of each relay is observed.

Table 4.11 Results obtained with proposed interval method for the microgrid operating in island mode.

No.	Relay	Primary time (Close-in)	Backup time	
			Bus-end	Remote line-end
1	23	0.177	0.417	0.763
2	24	0.138	0.497	0.730
3	36	0.194	0.458	0.718
4	46	0.107	0.404	0.767
5	51	0.100	0.402	0.773
6	32	0.170	0.417	0.723
7	42	0.192	0.406	0.701
8	63	0.186	0.406	0.749
9	64	0.115	0.487	0.742
10	65	0.120	0.401	0.725
Number of Violations (NV)				0
Computation Time (s)				15.21

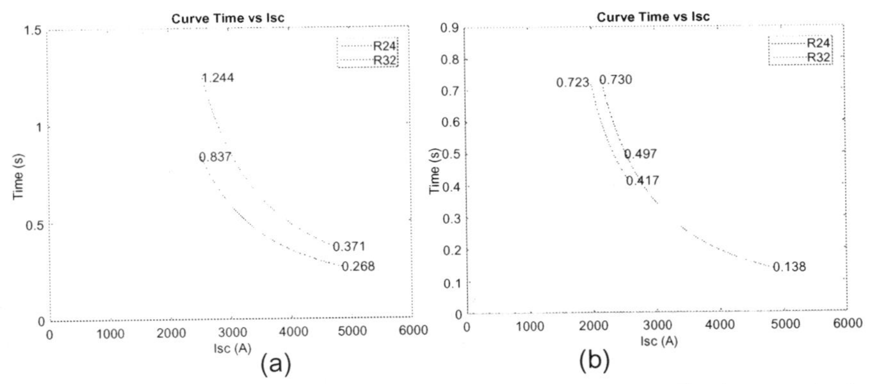

Fig. 4.34 Relay time operation comparison between coordination methods: (A) Conventional method, (B) Interval method.

5.4 Contingency analysis on the 14 IEEE bus

The evaluation of the response of the coordination method to contingencies that are subject to operational or topological changes is important to determine the robustness of the relay coordination. The purpose is to understand the variation in the number of violated relays and the increase in the operation times of relays in the face of dynamic changes in the electrical network. With this, we will be able to know the resilience that the protection scheme presents.

In this case, outline 23 and generator G-2 are carried out like contingencies; the relay coordination changes due to the fault current variations and load current. The variation in I_{sc} seen by each of the relays under these conditions is evaluated in Fig. 4.35. As can be seen, there is a decrease in the I_{sc} seen by the relays, which may be higher or lower in some cases depending on the network topology.

Table 4.12 shows the results obtained by applying the conventional coordination method in the event of contingencies; as can be seen, there are six violations or losses of coordination, in addition to certain cases where coordination between the primary element and the backup element is

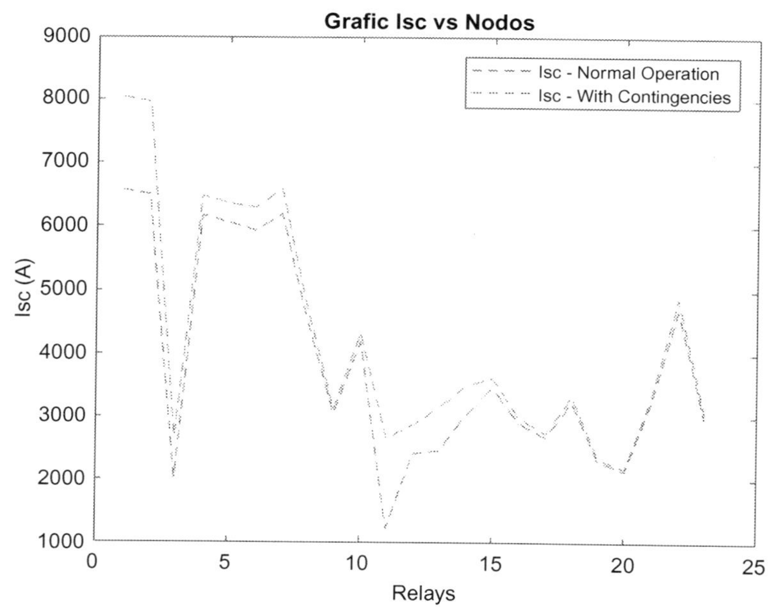

Fig. 4.35 Fault current seen by the relays during normal operating conditions and contingencies (outline 23 and G-2).

Table 4.12 Results obtained with the conventional method for the IEEE system of 14 buses in contingencies (outline 23 and G-2).

Relay					Relay				
Primary	Backup	t_p	t_b	CTI_c	Primary	Backup	t_p	t_b	CTI_c
24	12	0.322	0.672	0.351	52	54	0.519	0.687	0.168
25	12	0.361	0.671	0.310	52	15	0.519	0.855	0.337
45	24	0.444	0.771	0.327	54	45	0.381	0.835	0.455
612	116	0.454	0.961	0.507	54	15	0.381	0.693	0.312
613	116	0.560	0.878	0.318	116	25	0.433	0.753	0.319
613	126	0.560	0.862	0.302	126	1011	0.166	0.917	0.751
910	149	0.411	0.726	0.315	136	1312	0.178	0.482	0.304
914	109	0.124	0.535	0.411	136	1213	0.178	0.634	0.456
1011	910	0.675	0.440	−0.235	109	1413	0.384	0.157	−0.227
1213	612	0.151	0.506	0.355	149	1110	0.383	0.188	−0.195
1314	613	0.549	0.862	0.313	1110	1314	0.139	0.540	0.400
1314	1213	0.549	0.164	−0.385	1312	611	0.465	0.828	0.363
21	52	2.991	3.295	0.304	1312	613	0.465	0.770	0.305
51	25	0.473	0.783	0.310	1413	1413	0.448	0.145	−0.304
42	45	0.189	0.523	0.334					
Number of Violations (NV)					6				
Computation Time (s)					DE: 37.21				

achieved. The resulting times are high and unproductive for protection purposes, as they can negatively influence the operation and stability of the network.

When applying the algorithm of coordination by intervals for the contingency scenario, adjustments are obtained that guarantee the operation of the relays in the established time interval (see Table 4.13). As can be seen, there is a decrease in the times of operation for faults in the primary and backup zone, by including a limitation in the operating time for faults at the remote end. The backups function is available, and the protection scheme times are within each specified interval, which positively influences the operation of the network.

5.5 Minimum demand scenario and sensitivity evaluation in the IEEE 14 bus system

With the settings defined in the normal operating state, the operating times are recalculated in a minimum demand scenario. For this, a decrease in I_{Load} between 30% and 50% is considered; the sensitivity of the backup elements to this condition is also evaluated. In Fig. 4.36A, the sensitivity of the backup

Table 4.13 Results obtained with the intervals propose method for the IEEE system of 14 buses in contingencies (outline 23 and G-2).

No.	Relay	Primary time (Close-in)	Backup time Bus-end	Backup time Remote line-end	No.	Relay	Primary time (Close-in)	Backup time Bus-end	Backup time Remote line-end
1	24	0.128	0.499	0.700	11	52	0.112	0.464	0.717
2	25	0.170	0.404	0.781	12	54	0.149	0.415	0.703
3	45	0.188	0.401	0.719	13	116	0.160	0.406	0.731
4	612	0.199	0.400	0.712	14	126	0.135	0.409	0.738
5	613	0.190	0.402	0.741	15	109	0.157	0.403	0.701
6	910	0.191	0.400	0.793	16	149	0.157	0.428	0.706
7	914	0.195	0.404	0.713	17	1110	0.179	0.400	0.748
8	1011	0.173	0.403	0.773	18	1312	0.195	0.459	0.701
9	1213	0.178	0.414	0.700	19	1413	0.184	0.456	0.704
10	1314	0.190	0.444	0.703					
Number of Violations (NV)					0				
Computation Time (s)					33.05				

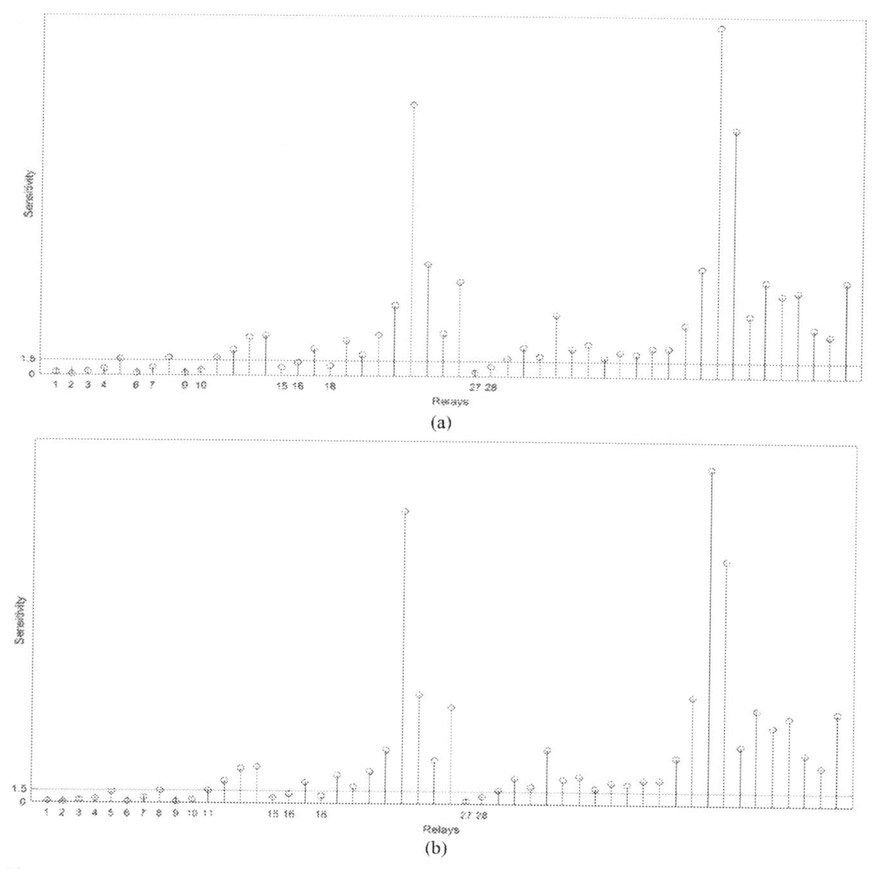

(a)

(b)

Fig. 4.36 Sensitivity of the backup relays: (A) Normal operation, (B) minimum demand.

relays that make up each of the coordination pairs of the IEEE 14 bus system under normal operation is shown; for this case, 13 relays do not meet the condition (>1.5), which satisfies a complete backup against faults at the remote end of the line.

The variation in the sensitivity of these elements is shown in the face of the minimum demand condition in the same IEEE bus 14 system; it can be seen that the same relays are maintained as in the previous conditions and others are added, reaching 16 cases that do not meet the sensitivity conditions (Fig. 4.36B). Table 4.14 includes a comparison of the primary and backup operating times for the normal and minimum demand operating conditions, as well as the coordination time intervals for each condition. Table E of the Appendix shows the values of the parameters and settings used for this analysis.

Table 4.15 shows the results of the analysis in minimum demand for the IEEE bus 14 system; in this case, the method is evaluated by intervals, the variation suffered by the operation times, and the coordination time interval for this scenario.

5.6 Weak source contribution

Under certain operating conditions, the contribution of sources may be limited, either by the type of source or the high impedance of the interconnection. This low contribution may not be detected by the relays losing sensitivity, maintaining a fault contribution that can cause sustained low voltages, and the blocking of fault line restoration actions, preventing it from reclosing. The impact of relay coordination due to low-current contribution of weak sources is evaluated. This operating condition is shown in Fig. 4.37, where the comparison of operating times between a relay (R_2) with a conventional curve and another with a nonconventional curve is presented. The main benefit is in the operating times and the increase in sensitivity for the detection of the fault condition. In Fig. 4.37B, the coordination graphs are shown with the time curves used and the times obtained. The implementation of dynamic load currents and in conjunction with nonconventional time curves allows the resilience of the relay to be significantly improved.

Table 4.14 Scenario of minimum demand for the 14-bus IEEE system, conventional method.

No.	Primary R	Backup R	t_p (100%)	t_p (50%)	t_b (100%)	t_b (50%)	*CTI* (100%)	*CTI* (50%)
1	23	12	0.178	0.165	0.660	0.428	0.482	0.264
2	24	12	0.248	0.237	0.658	0.428	0.411	0.191
3	25	12	0.339	0.329	0.657	0.428	0.317	0.099
4	34	23	0.321	0.260	0.622	0.274	0.301	0.014
5	45	24	0.333	0.228	0.641	0.336	0.308	0.107
6	45	34	0.333	0.228	0.643	0.338	0.310	0.109
7	612	116	0.663	0.644	0.971	0.478	0.309	−0.166
8	613	116	0.548	0.476	0.887	0.458	0.339	−0.018
9	613	126	0.548	0.476	0.869	0.298	0.320	−0.178
10	910	149	0.519	0.512	0.885	0.490	0.366	−0.023
11	914	109	0.172	0.167	0.607	0.443	0.435	0.276
12	1011	910	0.226	0.214	0.630	0.540	0.404	0.326
13	1213	612	0.439	0.434	0.745	0.665	0.306	0.231
14	1314	613	0.215	0.201	0.785	0.534	0.570	0.333
15	1314	1213	0.215	0.201	0.472	0.442	0.257	0.242
16	21	52	1.122	0.448	1.444	0.452	0.323	0.004
17	51	25	0.405	0.257	0.811	0.445	0.406	0.189
18	51	45	0.405	0.257	0.708	0.317	0.303	0.060
19	32	43	1.989	0.800	0.343	0.265	−1.645	−0.536
20	42	34	0.227	0.176	0.605	0.328	0.378	0.152
21	42	54	0.227	0.176	0.533	0.326	0.306	0.150
22	52	15	0.201	0.176	0.673	0.303	0.472	0.127
23	52	45	0.201	0.176	0.546	0.279	0.345	0.103
24	43	24	0.254	0.243	0.697	0.349	0.443	0.106
25	43	54	0.254	0.243	0.567	0.334	0.313	0.091
26	54	15	0.361	0.283	0.676	0.304	0.315	0.020
27	54	25	0.361	0.283	0.679	0.414	0.318	0.130
28	116	1011	0.439	0.345	0.297	0.232	−0.143	−0.113
29	126	1312	0.156	0.133	0.491	0.462	0.335	0.329
30	136	1213	0.147	0.098	0.453	0.438	0.306	0.340
31	136	1413	0.147	0.098	0.650	0.461	0.503	0.362
32	109	1110	0.436	0.401	0.195	0.122	−0.241	−0.279
33	149	1314	0.498	0.396	0.270	0.215	−0.227	−0.181
34	1110	611	0.114	0.102	0.657	0.533	0.543	0.431
35	1312	613	0.454	0.452	0.756	0.527	0.301	0.074
36	1312	1413	0.454	0.452	0.778	0.492	0.323	0.040
37	1413	914	0.461	0.413	0.196	0.173	−0.265	−0.240

Table 4.15 Scenario of minimum demand for the 14-bus IEEE system, intervals method.

No.	Relay	Ta (100%)	Tb (100%)	Ta (50%)	Tb (50%)	*CTI* (100%)	*CTI* (50%)
1	23	0.155	0.423	0.140	0.225	0.268	0.085
2	24	0.114	0.495	0.103	0.196	0.381	0.093
3	25	0.134	0.403	0.095	0.234	0.269	0.139
4	34	0.103	0.415	0.076	0.163	0.312	0.087
5	45	0.170	0.407	0.097	0.160	0.237	0.063
6	612	0.191	0.400	0.105	0.197	0.209	0.092
7	613	0.189	0.400	0.094	0.172	0.211	0.078
8	910	0.143	0.400	0.097	0.221	0.257	0.124
9	914	0.194	0.411	0.117	0.207	0.217	0.089
10	1011	0.184	0.401	0.108	0.213	0.217	0.105
11	1213	0.194	0.423	0.137	0.223	0.229	0.086
12	1314	0.196	0.443	0.117	0.195	0.247	0.078
13	52	0.161	0.402	0.116	0.229	0.241	0.114
14	43	0.153	0.463	0.093	0.198	0.310	0.104
15	54	0.192	0.408	0.083	0.143	0.216	0.061
16	116	0.138	0.439	0.062	0.147	0.301	0.085
17	126	0.103	0.403	0.071	0.194	0.300	0.123
18	109	0.176	0.400	0.094	0.190	0.224	0.096
19	149	0.191	0.448	0.121	0.186	0.257	0.065
20	1110	0.126	0.402	0.069	0.159	0.276	0.090
21	1312	0.125	0.401	0.090	0.206	0.276	0.116
22	1413	0.175	0.492	0.107	0.211	0.317	0.103

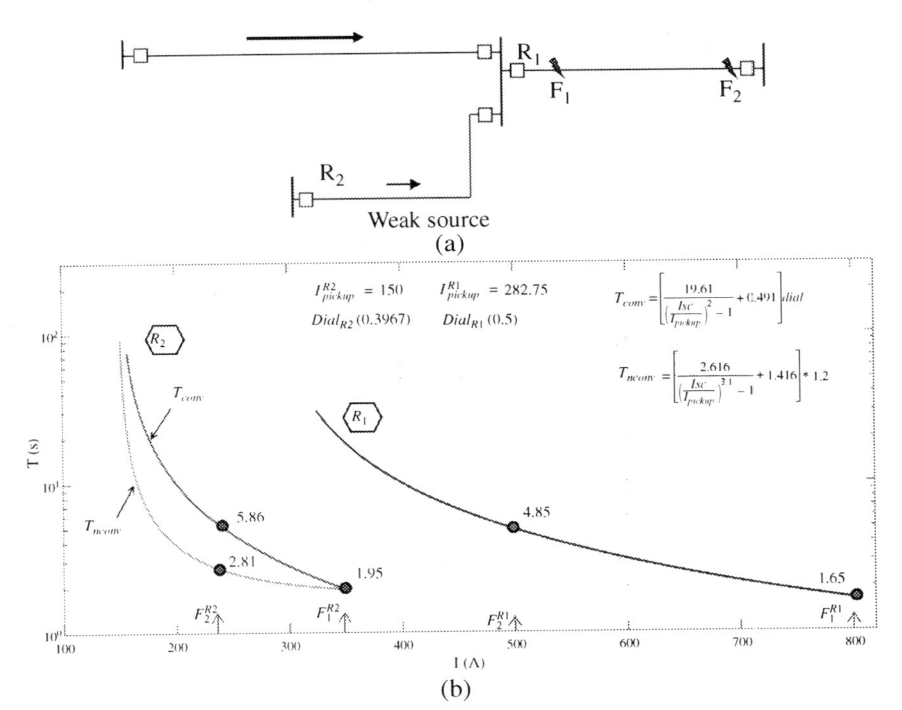

Fig. 4.37 Re-coordination of RDB for fault in line BC. (A) Electrical network, (B) coordination chart.

6. Conclusion

Active low-voltage networks represent a challenge for protection systems, the high topological dynamics and the presence of RAS compromise sensitivity, coordination, and relay operating times. Traditional protection schemes hardly meet the requirements for safety and speed in their operation. On-line protection adjustment schemes supported with real-time measurements through micro phasers and with low latency communication channels support allows the implementation of adaptive schemes where protection settings are determined based on the current operating conditions of the power grid. Due to the complexity of the coordination of DOCRs and to satisfy the different operating scenarios, the implementation of nonstandardized curves and new coordination methods that do not have functional dependence between the overcurrent relays presents adequate characteristics for the protection of electrical systems, favoring the resilience of power grids subject to highly diverse operating conditions.

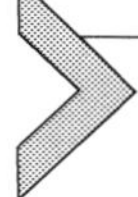

Appendix

Power system data

Table A Power flow and fault currents for 14-bus IEEE system in normal operation.

Relay		I_{sc}		I_{Load}	
Primary	**Backup**	**Primary**	**Backup**	**Primary**	**Backup**
23	12	8103	5688	292	620
24	12	8035	5698	225	620
25	12	7953	5711	169	620
34	23	1640	1647	104	292
45	24	2667	1543	250	225
45	34	2667	784	250	104
612	116	6481	1336	136	196
613	116	6356	1425	327	196
613	126	6356	592	327	136
910	149	6297	1118	82	144
914	109	6581	1092	144	82
1011	910	4723	1723	140	82
1213	612	3124	3124	38	136
1314	613	4325	3452	134	327
1314	1213	4325	1283	134	38
21	52	2656	984	620	169

51	25	2876	1377	302	169
51	45	2876	1420	302	250
32	43	1410	1425	292	104
42	34	3148	784	225	104
42	54	3148	2242	225	250
52	15	3490	1702	169	302
52	45	3490	1700	169	250
43	24	3686	1451	104	225
43	54	3686	2116	104	250
54	15	3619	1698	250	302
54	25	3619	1605	250	169
116	1011	2987	1987	196	140
126	1312	2730	1230	136	38
136	1213	3330	1283	327	38
136	1413	3330	1556	327	134
109	1110	2345	1345	82	140
149	1314	2173	2173	144	134
1110	611	3306	3306	140	196
1312	613	4880	3609	38	327
1312	1413	4880	1274	38	134
1413	914	3061	3061	134	144

Table B Power flow and fault currents for 14-bus IEEE system with DG penetration.

Relay		I_{sc}		I_{Load}	
Primary	**Backup**	**Primary**	**Backup**	**Primary**	**Backup**
23	12	8669	5716	292	620
24	12	8507	5729	225	620
25	12	8393	5744	169	620
34	23	1662	1669	104	292
45	24	3104	1580	250	225
45	34	3104	765	250	104
611	126	8501	759	196	136
611	136	8501	2965	196	327
612	116	10,715	2898	136	196
612	136	10,715	3276	136	327
613	116	9303	3018	327	196
613	126	9303	1737	327	136
910	149	8123	2630	82	144
914	109	8185	2802	144	82
1011	910	5676	1976	140	82
1213	612	3843	3843	38	136
1314	613	7816	4065	134	327
1314	1213	7816	1040	134	38

21	52	3236	846	620	169
51	25	3818	1401	302	169
51	45	3818	1798	302	250
32	43	1522	1538	292	104
42	34	4001	794	225	104
42	54	4001	2587	225	250
52	15	4406	1705	169	302
52	45	4406	2076	169	250
43	24	4553	1469	104	225
43	54	4553	2463	104	250
54	15	4138	1705	250	302
54	25	4138	1675	250	169
116	1011	6043	3316	196	140
126	1312	3688	3688	136	38
136	1213	7088	2033	327	38
136	1413	7088	2407	327	134
109	1110	3687	3687	82	140
149	1314	5496	2794	144	134
1110	611	6748	4030	140	196
1312	613	9323	4378	38	327
1312	1413	9323	2222	38	134
1413	914	6056	3343	134	144

Table C Power flow and fault currents for MG system connected to the utility.

Relay		I_{sc}		I_{Load}	
Primary	**Backup**	**Primary**	**Backup**	**Primary**	**Backup**
23	42	5504.49	3216	878	729
24	32	5316.16	3069	729	878
36	23	5230.72	2981	599	878
46	14	6166.69	3536	681	1185
46	24	6166.69	2810	681	729
56	15	5983.62	3778	839	1024
41	24	5646.76	2526	1185	729
41	64	5646.76	3247	1185	681
51	65	5839.97	3553	1024	839
61	36	7216.93	2332	1706	599
61	46	7216.93	2314	1706	681
61	56	7216.93	2732	1706	839
32	63	5828.59	3540	878	599
42	14	5847.79	3003	729	1185
42	64	5847.79	3031	729	681
63	46	6994.82	2396	599	681
63	56	6994.82	2302	599	839

64	36	7674.6	2628	681	599
64	56	7674.6	2450	681	839
65	36	7194.92	2297	839	599
65	46	7194.92	2116	839	681

Table D Power flow and fault currents for MG system operating in island mode.

Relay		I_{sc}		I_{Load}	
Primary	**Backup**	**Primary**	**Backup**	**Primary**	**Backup**
16	51	4098	2158	1706	1024
23	42	4496	2168	878	729
24	32	4852	2558	729	878
36	23	4798	2503	599	878
46	24	4274	2529	681	729
41	24	4462	2323	1185	729
41	64	4462	2239	1185	681
51	65	4701	2371	1024	839
61	36	5139	2141	1706	599
61	56	5139	1738	1706	839
32	63	4723	2395	878	599
42	64	3439	2165	729	681
63	46	3981	1659	599	681
63	56	3981	1554	599	839
64	36	4977	2347	681	599
64	56	4977	1661	681	839
65	36	4714	2146	839	599

Table E I_{Load} and sensitivity for the 14 IEEE bus in normal operation and minimum demand scenarios.

		Normal operation		Min demand				Normal operation		Min demand	
No.	Backup relay	I_{Load}	Sensitivity	I_{Load}	Sensitivity	No.	Backup relay	I_{Load}	Sensitivity	I_{Load}	Sensitivity
1	51	302	0.24	152	0.18	26	1213	38	9.09	19	10.74
2	21	620	0.07	337	0.05	27	32	292	0.35	149	0.32
3	42	225	0.37	115	0.31	28	42	225	0.95	115	0.87
4	52	169	0.61	85	0.51	29	52	169	1.73	85	1.58
5	12	620	1.58	337	1.34	30	25	169	2.80	85	2.93
6	32	292	0.19	149	0.16	31	45	250	1.95	126	2.02
7	52	169	0.76	85	0.65	32	43	104	5.96	52	6.11
8	12	620	1.70	337	1.45	33	34	104	2.69	52	2.84
9	32	292	0.19	149	0.16	34	54	250	3.14	126	3.17
10	42	225	0.51	115	0.43	35	15	302	1.85	152	1.84
11	12	620	1.74	337	1.47	36	45	250	2.37	126	2.46
12	23	292	2.48	149	2.55	37	24	225	2.21	115	2.31
13	24	225	3.74	115	3.92	38	54	250	2.79	126	2.79
14	34	104	3.88	52	4.11	39	15	302	2.77	152	2.79
15	126	136	0.79	69	0.67	40	25	169	5.02	85	5.26
16	136	327	1.28	164	1.09	41	1011	140	10.45	70	11.97
17	116	196	2.63	100	2.41	42	1312	38	33.96	19	37.02
18	136	327	0.98	164	0.91	43	1213	38	23.95	19	26.93
19	116	196	3.42	100	3.23	44	1413	134	5.94	68	6.53
20	126	136	2.05	69	1.95	45	1110	140	9.19	70	10.57
21	149	144	3.96	72	3.63	46	1314	134	7.90	68	8.68
22	109	82	6.82	41	6.00	47	611	196	8.21	100	9.67
23	910	82	26.24	41	32.23	48	613	327	4.70	164	5.67
24	612	136	10.79	69	12.15	49	1413	134	4.04	68	4.28
25	613	327	4.14	164	4.89	50	914	144	9.20	72	10.26

References

Bedekar, P. P., Bhide, S. R., & Kale, V. S. (2011). Determining optimum TMS and PS of overcurrent relays using linear programming technique. In *The 8th electrical engineering/electronics, computer, telecommunications and information technology (ECTI) association of Thailand—Conference*. https://doi.org/10.1109/ECTICON.2011.5947936.

C37.112-2018. (2019). *IEEE standard for inverse-time characteristics equations for overcurrent relays*. https://doi.org/10.1109/IEEESTD.2019.8635630.

Carlos, A., Castillo, A. C., & Schaeffer, S. E. (2015). Directional overcurrent relay coordination considering non-standardized time curves. *Electric Power Systems Research, 122*, 42–49. https://doi.org/10.1016/j.epsr.2014.12.018.

Chowdhury, S., Ten, C. F., & Crossley, P. (2008). Islanding protection of distribution systems with distributed generators: A comprehensive survey report. In *IEEE power and energy society general meeting* (pp. 1–8). https://doi.org/10.1109/PES.2008.4596787.

Coffele, F., Booth, C., & Dyśko, A. (2015). An adaptive overcurrent protection scheme for distribution networks. *IEEE Transactions on Power Delivery, 30*(2), 561–568. https://doi.org/10.1109/TPWRD.2013.2294879.

Dehghanian, P., Wang, B., & Tasdighi, M. (2019). New protection schemes in smarter power grids with higher penetration of renewable energy systems. *Pathways to a Smarter Power System*, 317–342. https://doi.org/10.1016/B978-0-08-102592-5.00011-9.

Ezzeddine, M., & Kaczmarek, R. (2008). Reduction of effect of coordination constraints in the linear optimization of operating times of overcurrent relays. In *MELECON 2008—The 14th IEEE Mediterranean electrotechnical conference* (pp. 707–712). https://doi.org/10.1109/melcon.2008.4618518.

Huchel, L., & Zeineldin, H. H. (2016). Planning the coordination of directional overcurrent relays for distribution systems considering DG. *IEEE Transactions on Smart Grid*, 7(3), 1642–1649. https://doi.org/10.1109/TSG.2015.2420711.

IEC 60255-151:2009. (2009). *Measuring relays and protection equipment—Part 151: Functional requirements for over/under current protection*.

IEEE 1547-2018. (2018). *IEEE standard for interconnection and interoperability of distributed energy resources with associated electric power systems interfaces, SASB/SCC21—SCC21—fuel cells, photovoltaics, dispersed generation, and energy storage*.

Kida, A. A., & Gallego Pareja, L. A. (2016). Optimal coordination of overcurrent relays using mixed integer linear programming. *IEEE Latin America Transactions, 14*(3), 1289–1295. https://doi.org/10.1109/TLA.2016.7459611.

Lua, Y., & Chung, J.-L. (2013). Detecting and solving the coordination curve intersection problem of overcurrent relays in subtransmission systems with a new method. *Electric Power Systems Research, 95*, 19–27. https://doi.org/10.1016/j.epsr.2012.08.009.

Memon, A. A., & Kauhaniemi, K. (2015). A critical review of AC microgrid protection issues and available solutions. *Electric Power Systems Research, 25*, 23–31. https://doi.org/10.1016/j.epsr.2015.07.006.

Mladenovic, S., & Azadvar, A. A. (2010). Sympathetic trip prevention by applying simple current relays. In *IEEE PES general meeting*. https://doi.org/10.1109/PES.2010.5589955.

Noghabi, A. S., Sadeh, J., & Mashhadi, H. R. (2010). Optimal coordination of directional overcurrent relays considering different network topologies using interval linear programming. *IEEE Transactions on Power Delivery, 25*(3), 1348–1354. https://doi.org/10.1109/TPWRD.2010.2041560.

Peças Lopes, J. A., Hatziargyriou, N., Mutale, J., Djapic, P., & Jenkins, N. (2007). Integrating distributed generation into electric power systems: A review of drivers, challenges and opportunities. *Electric Power Systems Research*, 77(9), 1189–1203. https://doi.org/10.1016/j.epsr.2006.08.016.

Pérez, G., Conde, A., & Gutiérrez, G. (2019). Dynamic state of BESS in microgrids. In *IEEE international conference on environment and electrical engineering and IEEE industrial and commercial power systems Europe (EEEIC/I&CPS Europe)*. https://doi.org/10.1109/EEEIC.2019.8783652.

Peterson, K., Ahmadiahangar, R., Shabbir, N., & Vinnal, T. (2019). Analysis of microgrid configuration effects on energy efficiency. In *IEEE 60th international scientific conference on power and electrical engineering of Riga Technical University (RTUCON)*. https://doi.org/10.1109/RTUCON48111.2019.8982341.

Razavi, S.-E., Rahimib, E., Javadi, M. S., Nezhad, A. E., Lotfi, M., Shafie-khah, M., & Catalão, J. P. S. (2019). Impact of distributed generation on protection and voltage regulation of distribution systems: A review. *Renewable and Sustainable Energy Reviews*, *105*, 157–167. https://doi.org/10.1016/j.rser.2019.01.050.

Shih, M. Y., Salazar, C. A. C., & Enríquez, A. C. (2015). Adaptive directional overcurrent relay coordination using ant colony optimisation. *IET Generation, Transmission & Distribution*, *9*(14), 1–10. https://doi.org/10.1049/iet-gtd.2015.0394.

Singh, M. (2017). Protection coordination in distribution systems with and without distributed energy resources—A review. *Protection and Control of Modern Power Systems*, *2*, 27.

So, C., Li, K. K., Lai, K. T., & Fung, K. Y. (1997). Application of genetic algorithm for overcurrent relay coordination. In *6th international conference on developments in power systems protection* (pp. 66–69). https://doi.org/10.1049/cp:19970030.

Stagg, G. W., & El-Abiad, A. H. (1968). *Computer methods in power system analysis*. http://books.google.de/books/about/Computer_methods_in_power_system_analysi.html.

Storn, R., & Price, K. (1997). Differential evolution—A simple and efficient heuristic for global optimization over continuous spaces. *Journal of Global Optimization*, *11*(4), 341–359.

Yousaf, M., Muttaqi, K. M., & Sutanto, D. (2018). Assessment of protective device sensitivity with increasing penetration of distributed energy resources. In *2018 Australasian universities power engineering conference (AUPEC)*. https://doi.org/10.1109/AUPEC.2018.8757966.

Zeineldin, H., El-Saadany, E., & Salama, M. (2006). Optimal coordination of overcurrent relays using a modified particle swarm optimization. *Electric Power Systems Research*, *76*(11), 988–995. https://doi.org/10.1016/j.epsr.2005.12.001. Elsevier.

CHAPTER FIVE

Smart grid stability prediction using genetic algorithm-based extreme learning machine

Fanidhar Dewangan[a], Monalisa Biswal[a], Bhaskar Patnaik[b], Shazia Hasan[c], and Manohar Mishra[d]

[a]Department of Electrical Engineering, National Institute of Technology, Raipur, India
[b]Biju Patnaik University of Technology, Rourkela, Odisha, India
[c]Department of Electrical and Electronics Engineering, Birla Institute of Technology & Science Pilani, Dubai Campus, Dubai, United Arab Emirates
[d]Department of Electrical and Electronics Engineering, FET, Siksha "O" Anusandhan University, Bhubaneswar, India

1. Introduction

Utilization of the best-in-class innovations in the field of communication and computing technology has developed most industries or sectors as more brilliant in terms of effectiveness, usefulness, and nature of administration, and so forth. Nonetheless, the electrical power system (EPS) (or the electric utility grid) has generally stayed nonmodernized when contrasted with other sectors. The reasons can be credited to the way that the grids have been designed and developed over the years. They have been built basically as a rigid mechanical system with sans thoughtfulness regarding the conceivable future need of repair or innovative redesigning. With thriving interest for power, there has been a huge expansion of grid infrastructure and manual control of it has been progressively troublesome and testing (Fadlullah & Kato, 2015). A certain degree of automation in monitoring and controlling has got introduced over the last couple of decades. In any case, it scarcely meets the new age requirements in terms of grid performance based on its reliability, stability, and resiliency, on the backdrop of radically changing grid topology and dynamics.

Although the traditional EPS stays burdened with the aforementioned mentioned insufficiencies, fast exhausting renewable energy sources (RES) and resulting advances in viable technological solution in identification and implementation of renewable power generation have brought about additional challenges. These RESs, which are connected to the EPS at

Electric Power Systems Resiliency
https://doi.org/10.1016/B978-0-323-85536-5.00011-4

low voltage, distribution level, turn the grid topology from a rigid centralized system to a distributed generation system. Unlike the centralized system, the power flow in the distributed generation system is bidirectional, resulting in serious protection issues such as relay miscoordination. Then, again, the accessibility of RESs likewise has given a fillip to the idea of microgrid technology, which by being able to cater to the local load demand helps increasing the system stability. Microgrids, being one of the key smart-grid enabler, are going to have increased presence in the modern EPS. Implementation of microgrid technology has its share of issues and challenges, a detailed account of which is presented in Patnaik et al. (2020). The path of modernization of the existing EPS toward a smart grid is very daunting, though the expected benefit far outweighs it. In a smart grid, the operation and control activities rely primarily on a robust and efficient communication network, which is so essential for gathering of useful information from inbuilt sensors with various system components located at different places. The enormous data thus collected is to be preprocessed before being inputted to intelligent computational algorithms which should be able to provide appropriate and accurate decisions in the fastest possible time. In fact, we are looking forward to an EPS which is heavily reliant on intelligent computational techniques working in tandem with fast-acting communication networks serving as an all-important support system. So, to say precisely, the present-day EPS is evolving as a cyber-physical systems (CPS). A strongly interlinked cyber technologies with physical power infrastructure is referred to as CPS in smart grid scenario.

NIST Cyber-Physical Systems website defines CPS as (NIST, n.d.):

> *Cyber-Physical Systems (CPS) comprise interacting digital, analog, physical, and human components engineered for function through integrated physics and logic.*

A more generalized definition is given in Rajkumar et al. (2010) as:

> *CPS can be characterized as physical and engineered systems whose operations are monitored, controlled, coordinated, and integrated by a computing and communicating core.*

The CPS enables the smart grid to deliver the expected functional features of it, such as (Alazab et al., 2020):

Savings of investment in capacity enhancement through effectual power management.

Enhanced system reliability and stability through adaptive and self-healing protection mechanisms.

System stability increase through meeting the local load demand by facilitating plug-in of DGs.
Ensuring distribution flexibility for effective load management.
Facilitating short-term and long-term power demand prediction.
Facilitating pollution-free environment through renewable energy sources.
Reduction in cost of price of power.

The above-mentioned requirements of a smart grid that ensures the stability of the network can be realized by adopting intelligent computing algorithms that makes for the mainstay of the cyber technology in a CPS. Applications of these intelligent computational algorithms, such as advanced optimization techniques, expert system, fuzzy logic, neural network, and deep neural network, are used to design schemes that are meant for control, management, protection, power trading, pre- and postdisaster management of the EPS to achieve increased power system resiliency, reliability, and robustness that help adding to the stability of the smart grid.

In this work, an optimized extreme leaning machine (ELM) is suggested to predict the stability of the industrial smart grid (SG). The performance of the proposed model is then compared with other contemporary machine learning (ML) models.

The steps involved in this work can be presented as follows:

1. Initially, the SG data is retrieved from the UCI-ML repository (Arzamasov et al., 2018) and processed through a preprocessing step such as normalization and label encoding.
2. A novel GA-based optimized extreme learning machine (ELM) is proposed for training and testing of the data.
3. This extracted SG dataset is than processed through the proposed optimized ELM model for training and testing purpose.
4. The performance of the proposed ML model is then compared with other contemporary ML and DL approaches using few key indexes, such as accuracy, precision, recall, and *F*-score.

2. Literature survey

A smart grid vastly relies on the accumulation of a huge amount of data procured from numerous sensors placed at different strategic locations of the EPS, fast and robust internet-based communication channels, and of course a fast-acting intelligent computation algorithm. It is imperative that handling and managing such huge data, protecting the communication channel from cyber infringements, and engaging a fast-acting computational technique for accurate prediction of the set objective is of utmost importance for the realization of a smart grid. Smart grid deployment thus involves huge complexity. The availability of advanced intelligent systems, with techniques such as ML, DL, reinforcement learning (RL), and deep reinforcement learning (DRL), SG realization is becoming feasible (D. Zhang et al., 2018).

In this aspect, Zerdoumi et al. (2018) illustrates the use of big data analysis along with intelligent computational models in resolving the issue of bulk data processing in a SG. Different possible applications of big data in SG context have been listed out in Ghorbanian et al. (2019). Covert data integrity assault (CDIA) poses a serious threat to the reliability and safety of smart grid functionalities, as they have the potential to infringe the traditional bad-data detectors used SG control points. For identification of intrusion by CDIAs, Ahmed et al. (2019) have presented an unsupervised machine learning-based algorithm working on unlabeled dataset, popularly known as "isolation forest." The generic security system for protection against CDIAs is usually a three-level construct. The three levels are in chronological order and are known as "Protection," "Intrusion Detection System (IDSs)," and "Alleviation." The first level uses communication shields and data safeguarding measures and majority of CDIAs get screened at this level. In case of the first level getting compromised, IDS which forms the second level detects the infringement and generates precautionary signals so that appropriate preventive actions can be initiated. There are abundant literatures on ML-based IDSs proposed in recent times (Ahmed et al., 2018a, 2018b; Esmalifalak et al., 2017; Fadlullah et al., 2011; Ozay et al., 2016; Wang et al., 2017; Y. Zhang et al., 2011). Fadlullah et al. (2011) and Y. Zhang et al. (2011) depict application of ML algorithms in recognizing suspicious user activities in SG communication networks. Ozay et al. (2016) depict the use of several ML algorithms for CDIA detection in the physical layer of a SG. Esmalifalak et al. (2017) have applied support vector machine (SVM) classifier in the above context. Wang et al. (2017) have proposed a ML-based model for detection of time synchronization assault (TSA). Ahmed et al. (2018b) have proposed an Euclidean-Distance-based ML model to predict the CDIAs. Ahmed et al. (2018a) have suggested a genetic algorithm (GA) combined SVM, for future selection and classification purposes, respectively, to detect the threats due to CDIAs in the communication channel.

The third level of defense, that is, alleviation, serves as a restoration mechanism, which restores the system operation after the CPS-assault identification is confirmed at the SG control point. Many optimization and intelligent techniques too are available for implementation in the alleviation process.

Zone-specific load forecasting is another vital exercise which needs precision predictive outcomes for stability of a smart grid and many ML techniques are available for such purposes. Analyzing archived data on weather, load variation pattern, and energy generation in a given specific region, these ML-based predictive algorithms make accurate forecasting of load demand for the specified regions (Chu, 2019; Vinayakumar et al., 2019). Vinayakumar et al. (2019) suggest that the deep neural network (DNN) model used for generation and load demand forecasting is better than the contemporary regression model. Zainab et al. (2020) have presented a distributed processing framework based on big-data for load demand forecasting. Dynamic pattern of energy consumption through household appliances is also a major deterrent in SG stability. Appropriate energy management techniques help address this issue. An ICT-based consumption prediction technique is proposed in Bassamzadeh and Ghanem (2017).

In addition to the above-discussed factors of smart grid instability, cost of power also plays a vital role in this aspect, requiring appropriate demand-side management to mitigate the issue to certain extent. In this context, a decentralized smart grid control model to conduct demand-side management is proposed in Wood (2020). The model relies on analysis carried on electricity price versus grid frequency variation. In addition, the authors have worked on an optimized data matching machine learning technique along with a transparent open box learning model to achieve dynamic smart grid stability.

From these discussions on factors affecting the stability of a smart grid and different possible intelligent computing-based techniques to mitigate them, it becomes apparent that there also lies the need to have a prediction model for smart grid stability itself. In this regard, Moldovan and Salomie (2019) have suggested a machine learning-based smart grid stability forecasting scheme. This method engages three types of genetic algorithms for future selection and four intelligent classifiers which includes the GBM algorithm. Alazab et al. (2020) have illustrated few DL-based models such as recurrent neural network (RNN), long short-term memory (LSTM), and gated recurrent unit (GRU) for smart grid stability prediction.

So far from the above brief literature survey, it can be concluded that the problem of smart grid stability does not have a single all-inclusive scheme of solution. Also, it can be inferred that smart grid stability prediction, as an area of research, is mired with scarcity of literature in this regard. In the present study, it is envisaged to design a novel GA-based ELM model for smart grid stability prediction. The proposed approach shows better results compared to the traditional ML techniques.

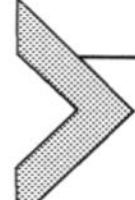

3. GA-based extreme learning machine (ELM)

3.1 Basic ELM

A basic ELM is essentially a fully connected single-hidden-layer feed forward neural network (SLFN) with random number of nodes in the hidden layer (Fig. 5.1). The input layer takes on randomly assigned weights and biases. The nonlinear activation function helps make the nonlinear input data linearly separable. The weights in the output layer are fixed by a noniterative linear optimization technique based on generalized-inverse. For a given set of n distinct training samples (x_i, t_i), x_i and t_i are described as:

$$x_i = [x_{i1}, x_{i2}, \ldots .x_{in}] \in R^n$$

and

$$t_i = [t_{i1}, t_{i2}, \ldots .t_{iL}] \in R^L$$

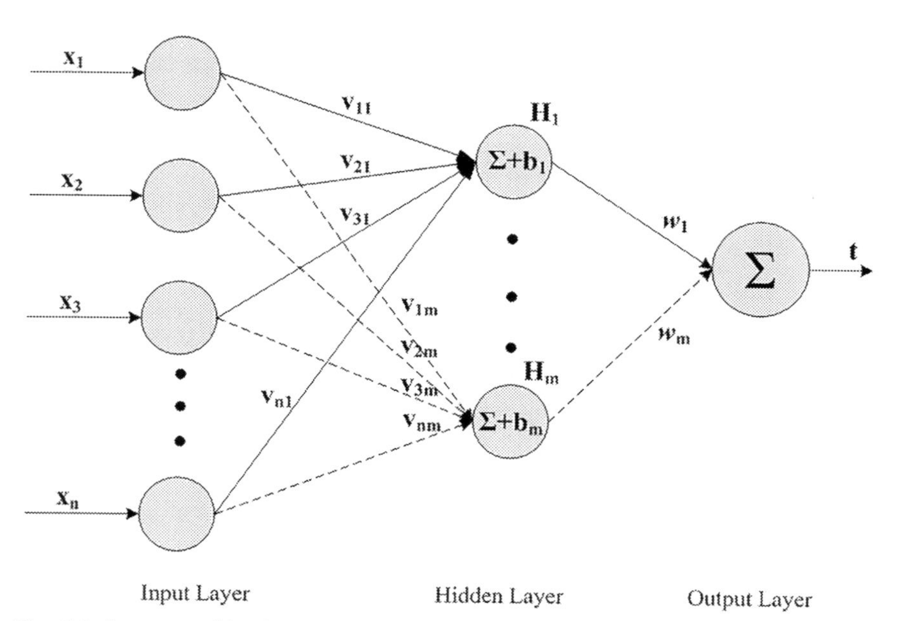

Fig. 5.1 Structure of basic ELM. *(No permission required.)*

where

n = dimension of input vector

m = number of hidden layer nodes

L = number of output classes

The output of kth hidden layer node (with $k < m$) is given as:

$$H_{ik} = \varphi\left(\sum\nolimits_{i=1}^{n} x_{ik} v_{ik} + b_k\right) \tag{5.1}$$

and the kth desired output is given as:

$$t_k = \sum\nolimits_{j=1}^{n} H_{jk} w_j \tag{5.2}$$

and

$$H \times w = t \tag{5.3}$$

where

$\varphi(.)$ is the activation function

$H = (h_{i1}, h_{12}, \ldots, h_{im})$ is the input layer weight matrix

$b = [b_1, b_2, \ldots, b_m]$ is the bias values of hidden layer neurons and the desired target output in the training set.

The hidden layer neurons approximate its input samples to be zero. But this approximation results in weaker generalization by the ELM due to over-fitting. This necessitates that the number of hidden layer nodes (m), which are chosen randomly/empirically, needs to be less than the number of input vector dimensions.

A pseudo-inverse of the matrix H, namely H^+, as H is not a full rank matrix, is determined by using a regularized least square solution as depicted in Eq. (5.4) below:

$$H^+ = \begin{cases} \left(H^T H + \lambda I\right)^{-1} H^T, & \text{for } L \leq N \\ \left(H H^T + \lambda I\right)^{-1} H^T, & \text{for } L > N \end{cases} \tag{5.4}$$

where λ is the regularization parameter.

However, for an accurate solution, the square matrix $(H^T H + \lambda I)$ needs to be invertible. And in SLFN as generally $m \leq n$, the matrix is mostly singular. The problem is addressed by Huang et al. (2006) by using SVD method. But gain, SVD suffers from low convergence and sluggish response for large data (Horata et al., 2013; Tzeng, 2013). In this context, a modified ELM is proposed in Yayık et al. (2019) using Hessenberg decomposition.

Hessenberg Decomposition ELM—HessELM:

The pseudo-inverse of the matrix H is obtained as

$$H^{+} = \left(HH^{T}\right)^{-1} H^{T} \tag{5.5}$$

wherein, HH^T is decomposed using Hessenberg decomposition.

$$HH^{T} = QU\,Q^{*} \tag{5.6}$$

where Q is an Unitary matrix and U being the upper Hessenberg matrix.

Substituting HH^T in Eq. (5.5);

$$H^{+} = \left(QUQ^{*}\right)^{-1} H^{T} = Q\,U^{-1}\,Q^{*}H^{T} \tag{5.7}$$

When considering singularity conditions, it is reached that $|U| \neq 0$ and U is nonsingular, and hence U^{-1} exists. The process to obtain output weights w in Eq. (5.3) and the optimum regularization parameter λ in Eq. (5.4) are enumerated in (Yayık et al., 2019). However, the basic ELM is used here for designing the proposed method and HessELM will be consider as a future scope for this work.

3.2 Genetic algorithm

Holland (1992) presents the essential standard of Genetic Algorithm (GA). GA is outstanding among other known optimization techniques broadly used to tackle complex problems (Bi, 2010; Mohamed, 2011). The point by point portrayal of the genuine coded GA is given in this section (refer Fig. 5.2).

Starting population

The first set of populace is a probable solution set P, which is a set of real values produced arbitrarily, and is given as:

$$P = \{p1, p2, \ldots, ps\}$$

Evaluation

A fitness function should be characterized to assess every chromosome in the populace, and the same is presented as:

$$\text{fitness} = g(P)$$

Selection

Once the fitness value is determined, the chromosomes are arranged based on their respective fitness values, and a widely popular "tournament selection method" is used for parent selection.

Genetic operators

After the selection process, certain genetic operators, such as "crossover" and "mutation" are used to build some new chromosomes (offspring) from

```
Basic GA implementation process

t ← 1                          //iteration number
P← initial populace comprising of set of random real values
g(P)                           //calculation of fitness value of each parent in P
while (condition: iteration not yet terminated) do
        t ← t+1
        O ← { }                //initialization of offspring populace
        while |O| < |P| do
                Selection of a pair of parents for crossover
                Mating of parents to create children O1 & O2
                O ← O U { O1, O2 }  and Mutate
        end
        P ← O and evaluate P, g(P)
end                            //new generation
```

Fig. 5.2 Basic GA implementation process. *(No permission required.)*

their parents (Zbigniew, 1996). The crossover operation (such as arithmetical crossover) involves exchange of information between two earlier selected parents. The offspring obtained after crossover operation are then subjected to mutation operation (such as uniform mutation operation), to change the genes of the chromosomes. After completion of the three successive operations of selection, crossover, and mutation, a new populace is obtained. The next iteration will begin with this new populace and the same process is repeated in the new iteration. This way, with every iteration, the best of the chromosomes of the present or new population gets selected and compared with the best of the same from the previous populace, ultimately leading to global best chromosome. The iterative process is stopped either when the results tend to converge or the number of iterations exceeds the stopping criteria.

3.3 Genetic algorithm-based ELM

The proposed GA methodology is applied to track down the optimal weights among input and hidden layers as well as bias values. The weights are bounded by [−1, 1], and the dimension of the populace relies on the number of input variables put to the ELM. At the beginning, the ELM is fed with the populace comprising of a set of random numbers, with classification accuracy considered as the fitness value for each of the chromosomes. The new populace thus generated is subjected to genetic iterations following the procedure as stated earlier using genetically optimized weights. After completion of the iterations, the proposed elitist genetic algorithm will result in optimized weights and bias.

4. Results and discussion

4.1 Dataset description

The data for the smart grid as is used in this experimentation has been picked up from the UCI-ML repository (Ahmed et al., 2018b). The dataset comprises of 10,000 samples and 14 features, out of which 12 are primary predictive features and the rest two are dependent features or variables.

These 12 predictive features help provide information, such as,

- reaction time (tau[x]) of smart-grid participants,
- nominal power ($p[x]$) consumed (negative)/produced (positive)
- co-efficient of elasticity of price ($g[x]$).

The dependent features on the other hand are given as:

- (stab): the maximal real part of the characteristic equation root (if +ve, the system is linearly unstable)
- (stabf): the stability label (class level) of the system (categorical: stable/ unstable).

4.2 Performance evolution matrices

This study has used the following evaluation indices for assessment of the performance of the proposed model through the obtained classification results.

Confusion matrix (CM): It reflects the actual values in respect of the predicted values (or number of samples) corresponding to different class levels in terms of a correct/true and incorrect/false prediction. As an example, Fig. 5.3 is provided as a demonstration of confusion matrix representation.

Eqs. (5.8)–(5.13) represent the performance indices formulated based on the above-cited confusion matrix.

Accuracy (A):

$$A = \frac{TS + TU}{TS + TU + FS + FU} \tag{5.8}$$

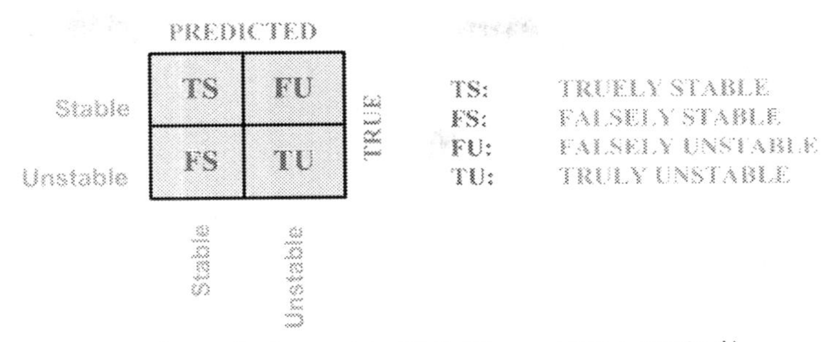

Fig. 5.3 Structure of a confusion matrix (CM). *(No permission required.)*

Precision (P):

$$P = \frac{TS}{TS + FU} \tag{5.9}$$

Recall (R):

$$R = \frac{TS}{TS + FS} \tag{5.10}$$

$F\beta$_Score:

$$F = \left(1 + \beta^2\right) \times \frac{R \times P}{\beta^2(R + P)} \tag{5.11}$$

$F1$_Score ($F1$): The score considers equal weight for both Precision as well as, and can be calculated by using β value of 1 in Eq. (5.11).

$$F1 = \frac{2RP}{R + P} \tag{5.12}$$

$F2$_Score ($F1$): It is the $F1$_Score calculated using $\beta = 2$, where precision carries less weight as compared to the recall value.

$$F2 = \frac{5RP}{4(R + P)} \tag{5.13}$$

4.3 Experimental results

In this section, a detailed performance evaluation of proposed GA-based ELM is carried out to predict the stability conditions of the smart grid. Initially, the total collected dataset is divided into two sets with a ratio of 80:20 (Training and Testing). The testing result in terms of CM is portrayed in Fig. 5.4.

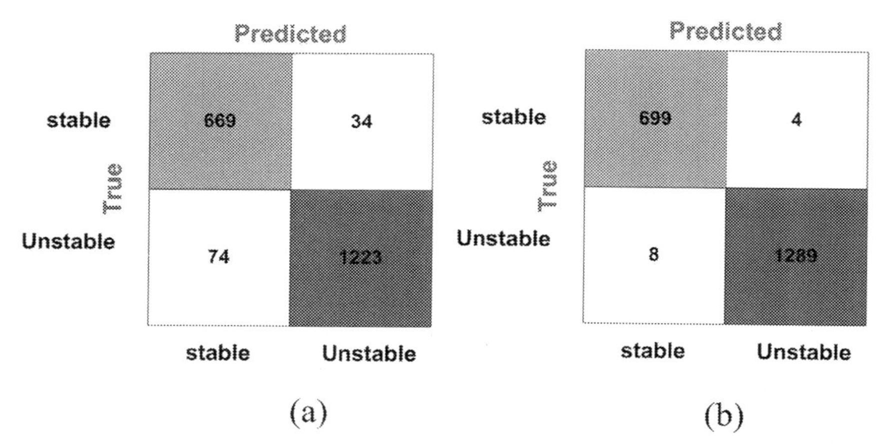

Fig. 5.4 Confusion matrix for (A) traditional ELM, (B) proposed GA-ELM. *(No permission required.)*

The performance result corresponding to the basic ELM and proposed GA_ELM are shown in Fig. 5.4A and Fig. 5.4B, respectively. It is clearly seen that the proposed method outperforms with less error compared with basic ELM in recognizing the stability of the smart grid. In case of GA-based ELM, only 18 stable cases are miss-classified compared with 34 cases in basic ELM. To get a clear picture of performance measure, several performance indices such as Precision, Recall, Accuracy, and Fβ-Score are calculated and tabulated in Table 5.1. We can clearly analyze from the table that the proposed GA_ELM model is able to classify the stability of the smart-grid data with 99.4% accuracy. The other indices such as precision, Recall, $F1$, $F2$, $F0.5$ scores, and area under the curve (AUC) are noted to be 0.994310, 0.988684, 0.9915, 0.6197, and 0.996086, respectively. It can be clearly seen from the table that the proposed model outperforms traditional ELM with respect to all the measuring indices.

The proposed prediction method is a hybrid model where the parameters of the ELM are tuned through GA. Fig. 5.5 displays the graph between Fitness (Accuracy) and number of generations. It has been seen that the proposed GA_ELM converses within 30 generations.

The performance of the proposed method is compared with few other contemporary ML techniques (such as decision tree, naïve bias, linear regression, random forest, and extreme gradient boosting) with similar dataset and training/testing ratio. The numerical results are presented in Table 5.1. Here, we can analyze that the performance of the naïve bias

Table 5.1 Performance metrics of prediction models.

Prediction models	Performance metrics					
	Precision	Recall	*F*1 Score	*F*2 Score	ROC-AUC	Accuracy (%)
DT	0.7488	1.0	0.8871	0.702315	0.61	74.88
NB	0.6485	1.0	0.786775	0.585075	0.5	64.85
LR	0.6485	1.0	0.786775	0.585075	0.5	64.85
RF	0.946227	0.909020	0.927251	0.5795	0.906857	90.75
XGBoost	0.910437	0.995373	0.951012	0.5944	0.907359	93.35
ELM	0.972951	0.942945	0.957713	0.5986	0.947290	94.6
Proposed model	0.994310	0.988684	0.9915	0.6197	0.996086	99.4

DT, decision tree; *LR*, linear regression; *NB*, naïve bias; *RF*, random forest; *XGBoost*, extreme gradient boosting.

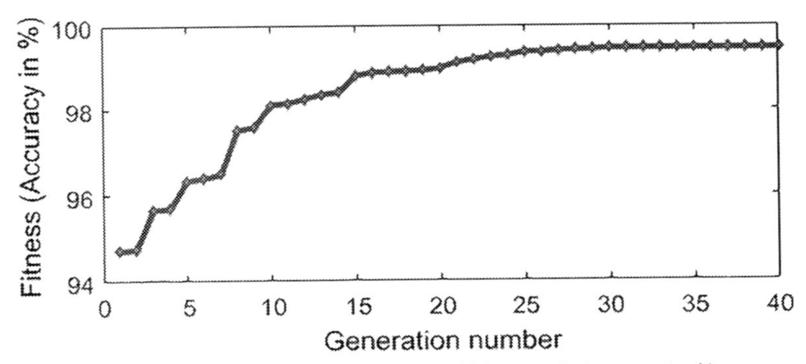

Fig. 5.5 Fitness changes in various generations. *(No permission required.)*

and linear regression provide extremely poor accuracy compared with decision tree, random forest, and extreme gradient boosting. Although the accuracy level of the XGBoost and basic ELM based approach has found to be 93.45% and 94.6%, respectively, it is much less than proposed GA_ELM-based approach (99.4%).

5. Conclusion

In this work, the smart-grid stability is predicted using an intelligent GA-based ELM approach to ensure the resiliency of the system. Here, 14-feature-based dataset extracted from UCI-ML repository is used for the validation purpose. The proposed optimized ELM technique provides an accuracy of 98%, which is better than basic ELM. The proposed approach is also compared with several other ML models with respect to the different performance measures. The comparative result shows that the proposed approach is better performer to classify the stability and instability of smart grid.

References

Ahmed, S., Lee, Y., Hyun, S. H., & Koo, I. (2018a). Covert cyber assault detection in smart grid networks utilizing feature selection and Euclidean distance-based machine learning. *Applied Sciences*, *8*(5), 772.

Ahmed, S., Lee, Y., Hyun, S. H., & Koo, I. (2018b). Feature selection-based detection of covert cyber deception assaults in smart grid communications networks using machine learning. *IEEE Access*, *6*, 27518–27529. https://doi.org/10.1109/ACCESS.2018.2835527.

Ahmed, S., Lee, Y., Hyun, S. H., & Koo, I. (2019). Unsupervised machine learning-based detection of covert data integrity assault in smart grid networks utilizing isolation forest. *IEEE Transactions on Information Forensics and Security*, *14*(10), 2765–2777. https://doi.org/10.1109/TIFS.2019.2902822.

Alazab, M., Khan, S., Krishnan, S. S. R., Pham, Q. V., Reddy, M. P. K., & Gadekallu, T. R. (2020). A multidirectional LSTM model for predicting the stability of a smart grid. *IEEE Access*, *8*, 85454–85463. https://doi.org/10.1109/ACCESS.2020.2991067.

Arzamasov, V., Klemens, B., & Jochem, P. (2018). Towards concise models of grid stability. In *2018 IEEE international conference on communications, control, and computing technologies for smart grids (SmartGridComm)* (pp. 1–6). https://doi.org/10.1109/SmartGridComm.2018.8587498.

Bassamzadeh, N., & Ghanem, R. (2017). Multiscale stochastic prediction of electricity demand in smart grids using Bayesian networks. *Applied Energy*, *193*, 369–380. https://doi.org/10.1016/j.apenergy.2017.01.017.

Bi, C. (2010). Deterministic local alignment methods improved by a simple genetic algorithm. *Neurocomputing*, *73*(13–15), 2394–2406. https://doi.org/10.1016/j.neucom.2010.01.023.

Chu, C. (2019). *Using machine learning to predict renewable energy generation and smart grid designs*. Princeton University Senior Theses. http://arks.princeton.edu/ark:/88435/dsp019g54xm49n.

Esmalifalak, M., Liu, L., Nguyen, N., Zheng, R., & Han, Z. (2017). Detecting stealthy false data injection using machine learning in smart grid. *IEEE Systems Journal*, *11*(3), 1644–1652. https://doi.org/10.1109/JSYST.2014.2341597.

Fadlullah, Z. M., Fouda, M. M., Kato, N., Shen, X., & Nozaki, Y. (2011). An early warning system against malicious activities for smart grid communications. *IEEE Network*, 25.

Fadlullah, Z. M., & Kato, N. (2015). *Evolution of smartgrids*. Springer International Publishing. https://doi.org/10.1007/978-3-319-25391-6.

Ghorbanian, M., Dolatabadi, S. H., & Siano, P. (2019). Big data issues in smart grids: A survey. *IEEE Systems Journal*, *13*(4), 4158–4168. https://doi.org/10.1109/JSYST.2019.2931879.

Holland, J. H. (1992). *Adaptation in natural and artificial systems: An introductory analysis with applications to biology, control, and artificial intelligence*. The MIT Press. https://doi.org/10.7551/mitpress/1090.001.0001.

Horata, P., Chiewchanwattana, S., & Sunat, K. (2013). Robust extreme learning machine. *Neurocomputing*, *102*, 31–44. https://doi.org/10.1016/j.neucom.2011.12.045.

Huang, G. B., Zhu, Q. Y., & Siew, C. K. (2006). Extreme learning machine: Theory and applications. *Neurocomputing*, *70*(1–3), 489–501. https://doi.org/10.1016/j.neucom.2005.12.126.

Mohamed, M. H. (2011). Rules extraction from constructively trained neural networks based on genetic algorithms. *Neurocomputing*, *74*(17), 3180–3192. https://doi.org/10.1016/j.neucom.2011.04.009.
Moldovan, D., & Salomie, I. (2019). Detection of sources of instability in smart grids using machine learning techniques. In *Proceedings - 2019 IEEE 15th international conference on intelligent computer communication and processing, ICCP 2019* (pp. 175–182). Institute of Electrical and Electronics Engineers Inc. https://doi.org/10.1109/ICCP48234.2019.8959649.
NIST. (n.d.). Retrieved from: https://www.nist.gov/el/cyber-physical-systems (Accessed 3 September 2021).
Ozay, M., Esnaola, I., Yarman Vural, F. T., Kulkarni, S. R., & Poor, H. V. (2016). Machine learning methods for attack detection in the smart grid. *IEEE Transactions on Neural Networks and Learning Systems*, *27*(8), 1773–1786. https://doi.org/10.1109/TNNLS.2015.2404803.
Patnaik, B., Mishra, M., Bansal, R. C., & Jena, R. K. (2020). AC microgrid protection—A review: Current and future prospective. *Applied Energy*, *271*, 115210.
Rajkumar, R., Lee, I., Sha, L., & Stankovic, J. (2010). Cyber-physical systems: The next computing revolution. In *Proceedings - Design automation conference* (pp. 731–736). https://doi.org/10.1145/1837274.1837461.
Tzeng, J. (2013). Split-and-combine singular value decomposition for large-scale matrix. *Journal of Applied Mathematics*, 2013. https://doi.org/10.1155/2013/683053.
Vinayakumar, R., Alazab, M., Soman, K. P., Poornachandran, P., Al-Nemrat, A., & Venkatraman, S. (2019). Deep learning approach for intelligent intrusion detection system. *IEEE Access*, 7, 41525–41550. https://doi.org/10.1109/ACCESS.2019.2895334.
Wang, J., Tu, W., Hui, L. C. K., Yiu, S. M., & Wang, E. K. (2017). Detecting time synchronization attacks in cyber-physical systems with machine learning techniques. In *Proceedings - International conference on distributed computing systems* (pp. 2246–2251). Institute of Electrical and Electronics Engineers Inc. https://doi.org/10.1109/ICDCS.2017.25.
Wood, D. A. (2020). Predicting stability of a decentralized power grid linking electricity price formulation to grid frequency applying an optimized data-matching learning network to simulated data. *Technology and Economics of Smart Grids and Sustainable Energy*, *5* (1). https://doi.org/10.1007/s40866-019-0074-0.
Yayık, A., Kutlu, Y., & Altan, G. (2019). *Regularized HessELM and inclined entropy measurement for congestive heart failure prediction.*
Zainab, A., Ghrayeb, A., Houchati, M., Refaat, S. S., & Abu-Rub, H. (2020). Performance evaluation of tree-based models for big data load forecasting using randomized hyperparameter tuning. In *Proceedings - 2020 IEEE international conference on Big Data, Big Data 2020* (pp. 5332–5339). Institute of Electrical and Electronics Engineers Inc. https://doi.org/10.1109/BigData50022.2020.9378423.
Zbigniew, M. (1996). Genetic algorithms + data structures = evolution programs. *Computational Statistics*, 372–373.
Zerdoumi, S., Sabri, A. Q. M., Kamsin, A., Hashem, I. A. T., Gani, A., Hakak, S., Al-garadi, M. A., & Chang, V. (2018). Image pattern recognition in big data: Taxonomy and open challenges: Survey. *Multimedia Tools and Applications*, 77(8), 10091–10121. https://doi.org/10.1007/s11042-017-5045-7.
Zhang, D., Han, X., & Deng, C. (2018). Review on the research and practice of deep learning and reinforcement learning in smart grids. *CSEE Journal of Power and Energy Systems*, *4* (3), 362–370. https://doi.org/10.17775/cseejpes.2018.00520.
Zhang, Y., Wang, L., Sun, W., Green, R. C., & Alam, M. (2011). Distributed intrusion detection system in a multi-layer network architecture of smart grids. *IEEE Transactions on Smart Grid*, *2*(4), 796–808. https://doi.org/10.1109/TSG.2011.2159818.

CHAPTER SIX

Enhancing relay resiliency in an active distribution network using latest data-driven protection schemes

Adhishree Srivastava[a] **and Sanjoy Kumar Parida**[b]
[a]Birla Institute of Technology, Mesra, Ranchi, India
[b]Indian Institute of Technology, Patna, Bihta, India

1. Introduction

The energy demand is expanding enormously day by day. To meet such power demands, use of distributed energy resources (DERs) is the most feasible solution. This has led to the formation of microgrids (MG), making present distribution grids active. The micro sources such as photovoltaic, wind turbine, and fuel cell generate power in an intermittent manner. Therefore, to maintain the power quality, voltage, and frequency stability, various semiconductor-based converters are required. These converters are of AC/AC, DC/AC, AC/DC, or DC/DC type which solely depends on type of load and source present in MG. These MG can operate either in standalone mode, where it is self-sufficient to supply all loads or in grid tied mode where large grid of infinite capacity is also incorporated with MG (Hatziargyriou, 2014).

Because of the presence of multiple sources and loads, protection of such grids is a big challenge. The protection system must be designed in such a way that it can respond to fault in both standalone and grid connected mode. If the fault is in grid side, required response is to detach the MG within permissible time. This is possible with the opening of a circuit breaker guided by directional overcurrent relay at the point of common coupling (PCC). Similarly, if fault presence on MG side is detected, protection system should isolate minimum portion of distribution feeders so that, fault is eliminated with the least load cut off. This may create sub microgrids where proper supply to each load, especially critical ones, must be assured.

Electric Power Systems Resiliency
https://doi.org/10.1016/B978-0-323-85536-5.00001-1

2. Microgrid protection challenges

The introduction of distributed generation (DG) in a power distribution network causes it to depart from its radial nature. The power flow changes from being unidirectional to being bidirectional (Bedekar & Bhide, 2011). This change is reflected in the values of short circuit fault currents and in their direction as well (Bedekar et al., 2009). To minimize the damage during such abnormalities, various protection systems are in place such as directional overcurrent relays which are economical. The selectivity and reliability of operation are achieved using proper relay settings to attain healthy relay coordination. Any MG has to operate with or without main grid, and in both, mode of operating system dynamics changes. Therefore, for proper designing of protection scheme, it is necessary to clearly understand some challenges involved, which are discussed in further sections (Hosseini et al., 2016).

2.1 Variation in short circuit level

The conventional distribution networks are usually supplied by sources of infinite capacity present at a distant location from the load center. Such grids are designed to sustain fault current up to defined magnitude. The rating of switchgear and circuit breakers is chosen based on maximum short circuit level. When distributed generation (DG) sources are installed near customer end, the distribution network becomes active. They start supplying the local loads and are self-capable of meeting the load demand even in the absence of utility grid. Therefore, both grid and DG sources collaborate with each other to supply required demand. In such case, if a fault occurs, both grid and DG will feed the fault current. This might increase or decrease the network short circuit level based on the location of fault and DG. The previous protection scheme designed for conventional grid may fail as they are mostly based on fault current level. On contrary, in island mode of operation, only DG contributes to the fault current which is limited, thus fault current level is significantly reduced. The overcurrent relay may not sense those faults as pickup current threshold is not crossed. Thus, a common overcurrent protection cannot protect the microgrid in both grid tied and standalone mode.

2.2 Bidirectional power flow

Most of the conventional grids were radial in nature, which means power flow was from grid source to load in one direction. Under a fault condition, overcurrent relays were designed to operate for particular current direction (Fig. 6.1). It is explained with an example of nine bus radial system.

In the above network, when DGs are installed at let's say at four buses, the same fault will be supplied by both grid and DGs (Fig. 6.2). In previous case, for the fault, only circuit breaker R3 will operate. It can be operated by nondirectional overcurrent relay. But after fault isolation, the successive loads 2 and 3 will also get affected. However, if an MG structure is used, then in case of fault, both R3 and R4 should operate. Both relays will sense

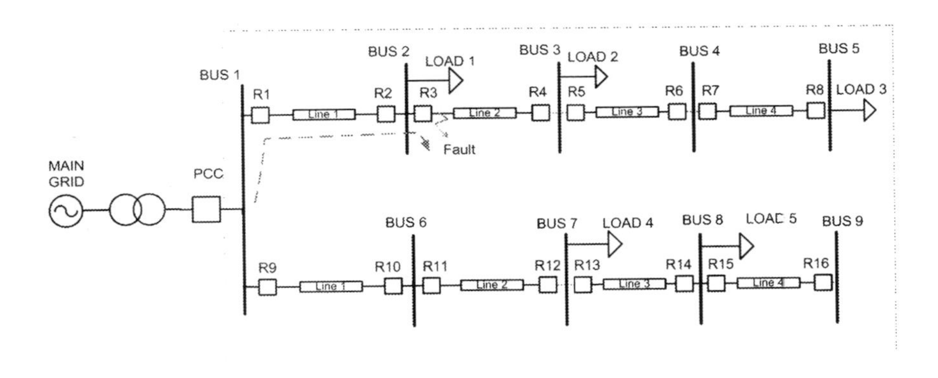

Fig. 6.1 Conventional radial distribution system. *(No permission required.)*

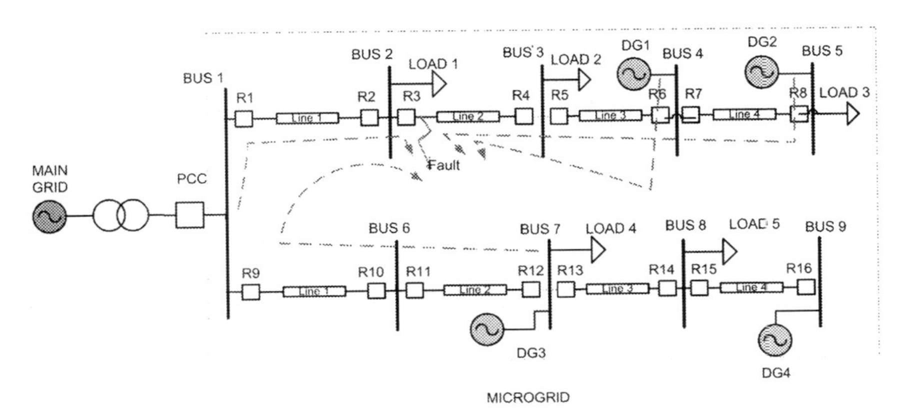

Fig. 6.2 Distribution grid changed to microgrid structure. *(No permission required.)*

fault because of the presence of DG source in downstream of fault. Thus, the relays must be of directional overcurrent type.

2.3 Fault current contribution based on DG type

Microgrids operate with two types of DGs, that is, synchronous based (conventional) or inverter based. Synchronous-based DGs like diesel generator, micro turbine, and gas turbine based are usually low inertia machine. Their output is sinusoidal and consistent. They can contribute up to five times the rated value (Zamani et al., 2010), and therefore, they can mislead the conventional protection scheme. Another type is inverter-based such as PV, fuel cell, battery storage, and wind turbine. They generate power at DC level or variable AC converted to constant DC, which are synchronized with distribution network through DC/AC converter. These converters consist of switches of limited current rating. Converters have inherent protection scheme of switches; therefore, maximum contribution to fault current is 1.5–2 times of rated value. Therefore, inverter-based DG contributes less toward fault current compared with synchronous DG.

2.4 Probabilistic power output of renewable-based DG

Power output from renewable energy sources (RES) like wind, solar irradiance, and tidal energy are intermittent in nature. Therefore, it is not assured that at the time of fault, RES DG contribution will be constant. In such case, continuous monitoring and control of RES DG output and load demand balance is necessary. Moreover, because of large number of converter present in MG, current and voltage harmonic prevails, which affects the protection scheme devised.

2.5 Unwanted tripping

It is possible that fault occurring in one feeder may cause tripping of subsequent feeder also if a high penetration of DG is available in its downstream. For example, as per Fig. 6.2, fault is present in line 2, but DG 1 is also contributing in fault current passing through line 3. If DG1 is highly penetrated, CB R5 and R6 may also trip isolating line 3, unnecessarily. Such unwanted tripping will cause power outage for more number of customers.

2.6 Blindness of protection

Pickup current is configured in a way that it is more than rated current of feeder and less than minimum short-circuit current of the protected zone.

When DG is installed to a network, the feeder equivalent impedance is increased and fault current may decrease. The relay in that zone may not sense the fault.

2.7 Possible conflict between feeder protection and fault ride through (FRT) requirement

As per the grid code and report presented in Grid code report, "GC0062–Fault ride through", National Grid (n.d.), a DG cannot disconnect itself from main distribution network on its own during a fault scenario. It must stay connected and bear that fault until it receives a disconnection command from the main control room. This capability of DG to stay connected for short time during fault is termed as fault ride through capability (FRT). The FRT sometimes contradicts with protection scheme and creates a confusion.

3. Protection techniques

As there are various protection issues in an MG, solving all issues with single protection scheme is the biggest challenge. However, a lot of researchers have proposed various ideas that can eradicate not all but most of the issues. Some of those protection schemes are discussed in further sections.

3.1 Short circuit level/overcurrent-based protection

The most effective protective device in any traditional distribution network is overcurrent (OC) relay. Its operation is simple and it is very economical. As discussed in challenges, with the addition of DG in conventional networks and their operation in islanded and grid-tied mode, short circuit level changes. Therefore, OC relay intended for traditional grid protection fails to operate when DG operation comes into picture. In that case, a reconfiguration of relay settings is necessary with addition of directional feature. An adaptive relay feature is required which can calculate the relay settings according to the network topology. Each time a DG is added or removed, the adaptive relay should update relay settings based on network situation. This can be done in offline or online manner. For online update of settings, a robust communication link, central protection computer, and digital relays are needed. Such protection is explained in next section.

3.2 Communication-assisted adaptive protection

In any MG system, at least two overcurrent relays must be installed at both ends of line feeder. Each DG unit is connected to distribution grid through a protection unit. Because a number of relays are present in such a system, their operation must coordinate with each other. For each protection zone, a combination of primary and backup relays is assigned. In case of fault, initially primary relay of dedicated zone should respond. If primary relay fails to operate because of some reasons, then only backup relay should operate. This means that backup relay must wait for primary relay to operate before activating itself. This specific time difference between primary and backup relay is called coordination time interval (CTI). The CTI between the backup and the primary relay depends on various parameters which are primary relay operating time, primary circuit breaker operating time, overshoot time, and signal travel time (refer Fig. 6.3). Its value lies typically between 0.2 and 0.5 s (Bedekar & Bhide, 2011). Therefore a relay coordination problem is formulated to find such relay settings that can ensure timely isolation of fault.

In literature, the most simple relay coordination problem is designed in the form of optimization problem which consists the minimization of single objective function. The overcurrent relays operate on varied operating characteristics, and for each characteristic, the time of operation is defined as:

$$t_{op} = \frac{\lambda(TMS)}{(PSM)^{\gamma} - 1} \tag{6.1}$$

$$t_{op} = \frac{\lambda(TMS)}{\left(I_{relay}/PS \times CT_{sec\ rated}\right)^{\gamma} - 1} \tag{6.2}$$

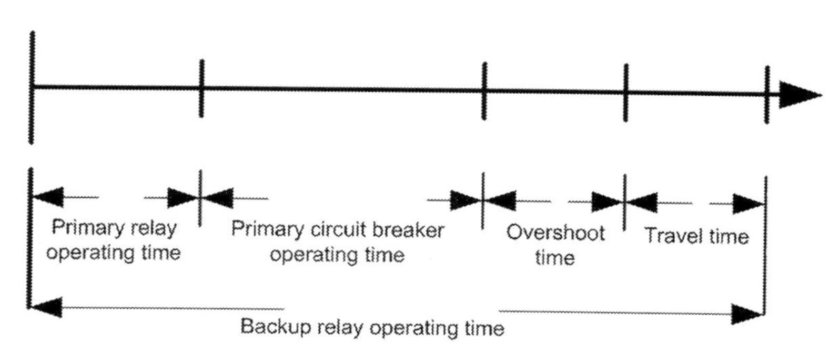

Fig. 6.3 Parameters for CTI.(Srivastava, Tripathi, Mohanty and Panda, 2016) *(No permission required.)*

Table 6.1 Value of λ and γ for various types of relays.

OC relay type	λ	γ
Instantaneous	Fixed operating time, no intentional time delay is added	
Definite time	Operating time is predefined and fixed time delay may be added if required	
Inverse definite minimum time (IDMT)	0.14	0.02
Very inverse	13.5	1
Extremely inverse	80	2

where t_{op} is the relay operating time, TMS is time multiplier setting, and PSM is plug setting multiplier. The constants λ and γ depend on the type of relay characteristics selected, and they are listed in Table 6.1 (Hosseini et al., 2016).

In this chapter, we will discuss the coordination problem assuming that each OC relays at feeder end are IDMT relay. The objective function of the problem is total operating time of all the relays present in the system. The function is to be minimized such that, each relay operates in minimum time and reliability of the system is maintained. The formulated objective function which is denoted here as "s" is:

$$min\ s = \sum_{i=1}^{n} t_{i,k} \tag{6.3}$$

where n is the number of relays, $t_{i,k}$ is operating time of ith relay for fault in kth zone. The constraints to solve this optimization problem are divided in three sections (Singh, 2013; Urdaneta et al., 1988).

Coordination criteria

$$t_{bi,k} - t_{i,k} \geq \Delta t \tag{6.4}$$

where $t_{i,k}$ is the operating time of primary relay at ith location for fault in zone k and $t_{bi,k}$ is the operating time of backup relay for fault in same zone, and Δt is the coordination time interval (CTI).

Bounds on relay operating time

$$t_{i,k\ min} \leq t_{i,k} \leq t_{i,k\ max} \tag{6.5}$$

where $t_{i,k\,min}$ is the minimum operating time of relay at *i*th location for fault in *k*th zone and $t_{i,k\,max}$ is the maximum operating time of the same. So, bound on time multiplier settings (TMS) will be

$$TMS_{i,k\ min} \leq TMS_{i,k} \leq TMS_{i,k\ max} \tag{6.6}$$

Its standard range usually lies between 0.025 and 1.2.

Bounds on pickup current

The minimum value of pickup current is determined by maximum load current seen by each relay. It is taken as 1.5 times of maximum load current so that normal and faulted condition can be easily differentiated. The maximum pickup current is determined by minimum fault current seen by each relay. This will impose bound on relay plug setting (PS) also, which is given below as:

$$I_{pmin} \leq I_p \leq I_{pmax}\ PS_{min} \leq PS \leq PS_{max} \tag{6.7}$$

Thus, there are two parameters of each relay, TMS and PS, which will function as design variable for this optimization problem satisfying every constraint. To solve this problem, a number of optimization techniques have been applied in literature. The fitness function mentioned above is a high constraint nonlinear, nonconvex optimization problem. Initially, the objective function was changed to linear function by taking a constant value of plug setting, and linear programming problem (LPP) is applied for OC relay coordination (Bhattacharya & Goswami, 2008; Urdaneta et al., 1988). By assuming a continuous value of PS and treating it as a variable, this nonlinear problem has been solved using nonlinear programming methods involving sequential quadratic programming method. Further, a number of search-based or heuristic optimization techniques has been applied to solve the given optimization problem such as genetic algorithm (GA), harmony search, particle swarm optimization (PSO), gravitational search algorithm (GSA), simulated annealing, and teaching learning algorithm in Mansour et al. (2007), Razavi et al. (2008), and Srivastava et al. (2016). To improve the results and get function convergence in minimum number of iteration, many hybrid optimization methods such as GA-NLP, GA-PSO, and PSOGSA are also applied (Sueiro et al., 2012).

After solving the above problem, a set of TMS and PS for each relay is obtained for grid connected as well as islanded mode. These settings confirm that a relay will operate only for the fault lying in its zone in minimum tolerant time while maintaining a proper time difference with backup relay. Now, as already discussed, MG can operate in grid-connected or islanded mode. Therefore the relay settings must be updated immediately when MG islands itself. To implement such operation, an adaptive protection scheme is required which can automatically detect the mode of MG operation and update the relay settings of each relays. To achieve such system, a highly efficient and rugged processor-based computer system is required that will be present in protection and control room of the grid area under consideration. All the relays must be replaced with numerical directional overcurrent relay (NDOCR) where the relay settings can be updated online through data received from remote or field intelligent electronic devices (IEDs) and communication links. The communication protocol of IEC 61850 can be adopted for this scheme (IEEE recommended practice for implementing an IEC 61850-based substation communications, protection, monitoring, and control system, Std, 2017). A simple adaptive system for MG of Fig. 6.1 is explained through Fig. 6.4.

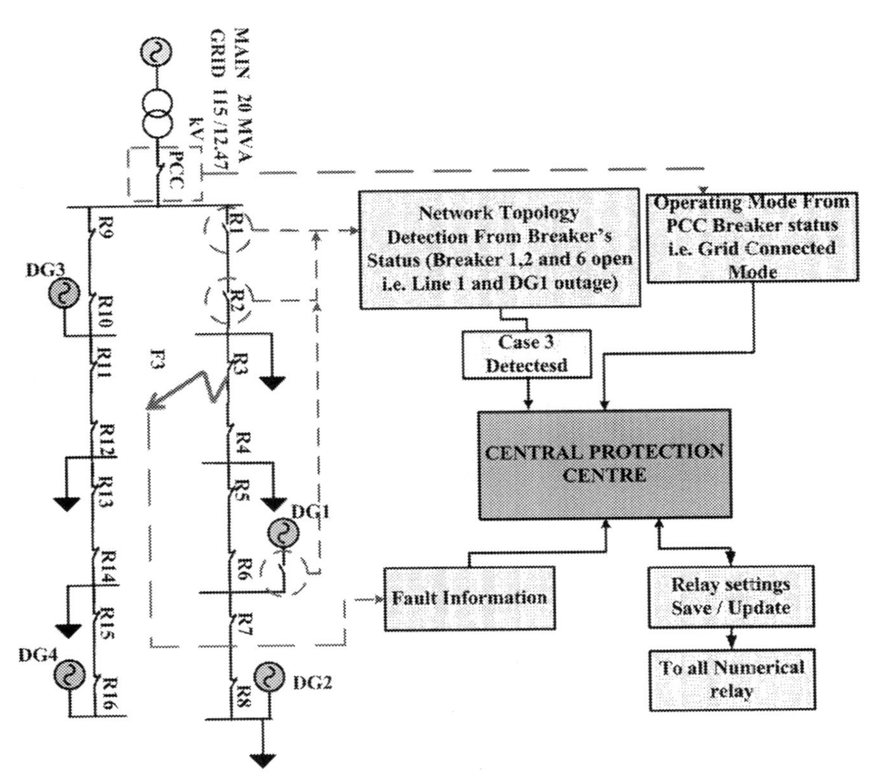

Fig. 6.4 Adaptive protection scheme.(Srivastava & Parida, 2020) *(No permission required.)*

In this example, the central protection center has following information:

1. Open/close status of each circuit breakers
2. Fault information from all current measurement devices
3. Relay settings stored for each network topology

All the information are transferred through the soft link shown with dashed line. The central protection center (CPC) continuously receives CB status from which it understands the present network topology of micro grid. The open status of point of common coupling (PCC) breaker corresponds to islanded state of MG and vice versa. When CPC receives the information of fault, it sends command to update relay settings based on the situation of MG topology. This online update completely relies on robust communication link. In case of link failure, it is important to preload the NDOCR with such relay settings that satisfies coordination constraints of all network topology.

3.3 Using fault current limiting device

One of the efficient methods to eradicate protection issues with DG system is to limit the short circuit level to a value such that protective relays could see similar value of fault current. In that case, relays can operate with a fixed pickup current setting and relay setting update is not required. However, because of dynamic nature of MG, location and size of fault current limiter is big issue. In case of fault, DGs output can be controlled or it can be disconnected. These methods basically aim at changing the fault current value seen by relay so that it can operate as per settings.

3.4 Advanced techniques

Based on ongoing research, an adaptive protection, with fast, reliable, and robust communication system, is supposed to be the most adequate solution for MG protection in both grid-tied and islanded mode. However, with this, a big risk may arise if the communication link fails. To combat that, an economical measure is proposed by Habib et al. (2018) where the energy storage device (ESD) starts contributing in fault current if command of PCC breaker opening is not received to relays. As discussed with islanding mode, lower relay settings are required but if those settings are not updated, the ESD will supply fault current such that small value of fault current reaches higher value, equivalent to the grid-tied mode. Seyedi and Karimi (2018) propose a comprehensive scheme of digital relays in which feeder relays coordinate with DG relays through more than one logic for relays. A digital relay triggering comes from an OR gate. This means those relays can be activated by more than one logic like:

- If overcurrent is observed in particular direction or
- If negative sequence current threshold is crossed in particular direction or
- If under voltage is observed or
- If total harmonic distortion (THD) of voltage and current supplied especially by inverters of MG crosses a minimum specified value.

Recently, some researchers have introduced machine learning (ML) application based on data-driven approach. This approach is being accepted by utilities due to development of phasor measurement unit (PMU). PMUs' ability to collect real-time magnitude and phase data of electrical quantities like current voltage and frequency assure availability of large data sets. Assuming the presence of PMU, in Seyedi and Karimi (2018), those data are processed through simple equation to detect and isolate fault in tolerant time of 0.2s. Machine learning algorithm, support vector machine (SVM), is applied in Agrawal and Thukaram (2013) where fault type and location information is received at central protection room by training a model of MG data. The three-phase current and voltage data are processed to ML model, which first detects whether fault is present or not, further it tells the type, location, and impedance of fault too. Another ML method GPR is applied in Srivastava and Parida (2020) for fault location prediction. A block representation is represented in Fig. 6.5.

Initially, the electrical signals received at generation buses of MG are retrieved and collected in database to use it as training data for ML model. After preprocessing of training data, it is used to design a robust ML model with the highest accuracy between actual and target output. Once the best ML model is selected in offline mode, it is ready to be installed for online mode. In online mode, the real-time electrical data during fault is sent as input to the installed ML model and it predicts the fault location and type. Various other authors have also proposed fault detection and classification using ML technique (Maruf et al., 2018; Mishra et al., 2019; Mishra & Rout, 2018; Abdelgayed, Morsi, & Sidhu, 2018; Srivastava & Parida, 2020). Thus, ML methods are a potential technique that can be applied for MG protection.

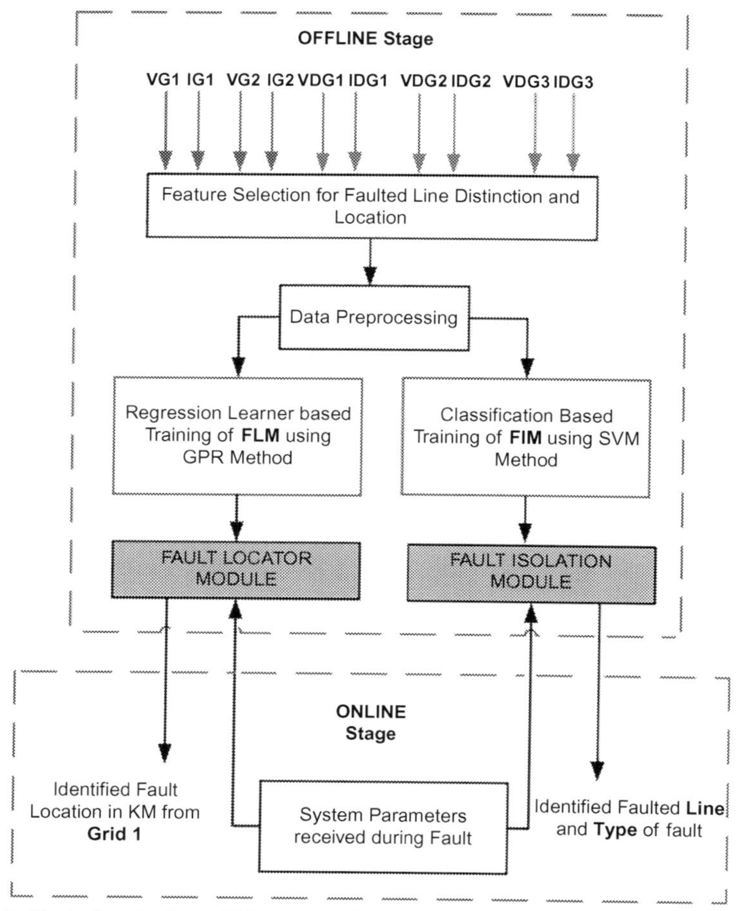

Fig. 6.5 Block description of ML approach (Srivastava & Parida, 2020). *(No permission required.)*

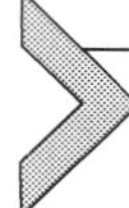

4. Comparative analysis of data-driven protection schemes

Resiliency of a micro grid is assessed by its ability to withstand any adverse scenario and how quickly it can bring the operation back to normal. A fault condition can occur in a micro grid anytime and as discussed in earlier sections, the challenges in micro grids render the conventional protection schemes to assure resiliency. The data-driven schemes consider every possible operating scenarios of a micro grid. Therefore as compared with conventional over current relays, a relay which is activated through data-driven decision-making is more reliable. A comparative analysis between conventional and data-driven protection scheme with respect to resiliency of an active distribution network can be summarized as below:

- Conventional over current relays sense current level to generate the trip commands but due to bidirectional nature of current and difference in short circuit level in an active network these relays cannot isolate fault with a fixed-relay setting. However, in data-driven approach, the relays are activated through another logic saved in each relay or available at central protection computer. The logic is generated after considering many electrical parameters of micro grid not just current level. Thus, such relays are resilient against faults and can reliably isolate faults in every operating condition of active network.
- Micro grid resiliency also considers timely restoration of power after fault has been isolated. This is possible if location of fault is identified quickly, so that personnel can be sent there for restoration. Conventional relays do not provide any information about exact location of fault. However, central protection computers trained with historical grid data using ML are capable of providing exact fault location just after it has been isolated.
- The active network data available at central protection system can also help in providing insights for operation and monitoring. The continuous data of various machines in a micro grid such as DGs, transformers, energy storage, and PV modules can help in performing predictive maintenance. The remaining useful life of these machines can be known before their actual failure. This can help in timely maintenance so that micro grid is healthy during operation and failure of machines could be the last reason for fault.

5. Conclusions

This chapter provides a brief overview of microgrid system, focusing mainly on its protection system. The basic challenges of MG operation are discussed through seven points, to cover almost all issues together. Further a number of solutions to overcome those challenges are also reviewed by classifying the protection techniques in four categories. The traditional protection scheme needs to be modified and advanced to save MG in both grid-connected and islanded mode. The adaptive relaying scheme with the aid of robust communication channel is becoming most promising way to overcome protection issues. Most recently, machine learning approaches have also arrived in this area of research, showing a high potential. This chapter will surely enhance reader's knowledge about current trends of MG protection, which can serve as a proper base for their further research.

References

Abdelgayed, T. S., Morsi, W. G., & Sidhu, T. S. (2018). Fault detection and classification based on co-training of semisupervised machine learning. *IEEE Transactions on Industrial Electronics*, *65*(2), 1595–1605. https://doi.org/10.1109/TIE.2017.2726961.

Agrawal, R., & Thukaram, D. (2013). Identification of fault location in power distribution system with distributed generation using support vector machines. In *2013 IEEE PES innovative smart grid technologies conference, ISGT 2013*. https://doi.org/10.1109/ISGT.2013.6497853.

Bedekar, P. P., & Bhide, S. R. (2011). Optimum coordination of directional overcurrent relays using the hybrid GA-NLP approach. *IEEE Transactions on Power Delivery*, *26*(1), 109–119. https://doi.org/10.1109/TPWRD.2010.2080289.

Bedekar, P. P., Bhide, S. R., & Kale, V. S. (2009). Coordination of overcurrent relays in distribution system using linear programming technique. In *2009 international conference on control automation, communication and energy conservation, INCACEC 2009*.

Bhattacharya, S. K., & Goswami, S. K. (2008). Distribution network reconfiguration considering protection coordination constraints. *Electric Power Components and Systems*, *36* (11), 1150–1165. https://doi.org/10.1080/15325000802084463.

Grid code report, "GC0062–Fault ride through", National Grid. (n.d.).

Habib, H. F., Lashway, C. R., & Mohammed, O. A. (2018). A review of communication failure impacts on adaptive microgrid protection schemes and the use of energy storage as a contingency. *IEEE Transactions on Industry Applications*, *54*(2), 1194–1207. https://doi.org/10.1109/TIA.2017.2776858.

Hatziargyriou, N. (2014). *Microgrids architectures and control*. Wiley, IEEE Press.

Hosseini, S. A., Abyaneh, H. A., Sadeghi, S. H. H., Razavi, F., & Nasiri, A. (2016). An overview of microgrid protection methods and the factors involved. *Renewable and Sustainable Energy Reviews*, *64*, 174–186. https://doi.org/10.1016/j.rser.2016.05.089.

IEEE recommended practice for implementing an IEC 61850-based substation communications, protection, monitoring, and control system, Std. (2017).

Mansour, M. M., Mekhamer, S. F., & El-Kharbawe, N. E. S. (2007). A modified particle swarm optimizer for the coordination of directional overcurrent relays. *IEEE Transactions on Power Delivery*, *22*(3), 1400–1410. https://doi.org/10.1109/TPWRD.2007.899259.

Maruf, H. M. M., Müller, F., Hassan, M. S., & Chowdhury, B. (2018). Locating faults in distribution systems in the presence of distributed generation using machine learning techniques. In *2018 9th IEEE international symposium on power electronics for distributed generation systems, PEDG 2018*Institute of Electrical and Electronics Engineers Inc. https://doi.org/10.1109/PEDG.2018.8447728.

Mishra, M., Panigrahi, R. R., & Rout, P. K. (2019). A combined mathematical morphology and extreme learning machine techniques based approach to micro-grid protection. *Ain Shams Engineering Journal*, *10*(2), 307–318. https://doi.org/10.1016/j.asej.2019.03.011.

Mishra, M., & Rout, P. K. (2018). Detection and classification of micro-grid faults based on HHT and machine learning techniques. *IET Generation, Transmission and Distribution*, *12* (2), 388–397. https://doi.org/10.1049/iet-gtd.2017.0502.

Razavi, F., Abyaneh, H. A., Al-Dabbagh, M., Mohammadi, R., & Torkaman, H. (2008). A new comprehensive genetic algorithm method for optimal overcurrent relays coordination. *Electric Power Systems Research*, *78*(4), 713–720. https://doi.org/10.1016/j.epsr.2007.05.013.

Seyedi, Y., & Karimi, H. (2018). Coordinated protection and control based on synchrophasor data processing in smart distribution networks. *IEEE Transactions on Power Systems*, *33*(1), 634–645. https://doi.org/10.1109/TPWRS.2017.2708662.

Singh, M. (2013). Protection coordination in grid connected & islanded modes of micro-grid operations. In *2013 IEEE innovative smart grid technologies - Asia, ISGT Asia 2013*. https://doi.org/10.1109/ISGT-Asia.2013.6698772.

Srivastava, A., & Parida, S. K. (2020). Fault isolation and location prediction using support vector machine and Gaussian process regression for meshed AC microgrid. In *2020 IEEE international conference on computing, power and communication technologies, GUCON 2020* (pp. 724–728). Institute of Electrical and Electronics Engineers Inc. https://doi.org/10.1109/GUCON48875.2020.9231261.

Srivastava, A., Tripathi, J. M., Mohanty, S. R., & Panda, B. (2016). Optimal over-current relay coordination with distributed generation using hybrid particle swarm optimization-gravitational search algorithm. *Electric Power Components and Systems*, *44*(5), 506–517. https://doi.org/10.1080/15325008.2015.1117539.

Srivastava, Adhishree, Parida, Sanjoy Kumar, et al. (2020). *Electric Power Components and Systems*, *48*(8), 781–798. https://doi.org/10.1080/15325008.2020.1821834. In press.

Sueiro, J. A., Diaz-Dorado, E., Míguez, E., & Cidrás, J. (2012). Coordination of directional overcurrent relay using evolutionary algorithm and linear programming. *International Journal of Electrical Power and Energy Systems*, *42*(1), 299–305. https://doi.org/10.1016/j.ijepes.2012.03.036.

Urdaneta, A. J., Nadira, R., & Pérez Jiménez, L. G. (1988). Optimal coordination of directional overcurrent relays in interconnected power systems. *IEEE Transactions on Power Delivery*, *3*(3), 903–911. https://doi.org/10.1109/61.193867.

Zamani, A., Sidhu, T., & Yazdani, A. (2010). A strategy for protection coordination in radial distribution networks with distributed generators. In *Proceedings of IEEE power and energy society general meeting* (pp. 1–8).

CHAPTER SEVEN

Microgrids as a resilience resource in the electric distribution grid

Gowtham Kandaperumal[a], Subir Majumder[b], and Anurag K. Srivastava[b]
[a]ComEd, Chicago, IL, United States
[b]Lane Department of Computer Science and Electrical Engineering, West Virginia University, Morgantown, WV, United States

1. Introduction

Recent statistics (Smith & Katz, 2013) show that the frequency of extreme weather events (also classified as low probability high impact, or LPHI, events), such as storms, floods, and wildfires, and associated substantiality of the power grid damage have soared up in the last decade. Recently occurred seven major storms have resulted into damages of over $1 billion each. While our power grid is designed to operate reliably, and industry-standard reliability indices such as SAIDI, SAIFI, CAIDI, and MAIFI (Khodayar et al., 2012) across multiple utilities around the globe strive to main associated countrywide-standard, these indices are primarily intended to capture low impact and high-frequency-type events. Continuing to be able to operate during events involving multiple contingencies is still a challenge. Furthermore, historical data shows that some of the areas are more prone to one kind of weather event than the other, and mitigation solutions to protect the grid for each kind of weather events can be different, and country-specific (Hussain et al., 2019). Therefore, there is no one-size-fits-all kind of solution available. Needless to say, the increasing severity of threats both from natural and weather events motivated researchers to develop a common measure to determine the impact of an event, but the globally accepted definition is still missing. However, as we will discuss later, criticality and uninteruptability of loads from the social point of view get higher precedence. It is intended that during these extreme events, the power utilities must able to be at least serve them.

Distribution networks are primarily the grid's load center. Increasing deployment of smart grid infrastructure and local distributed generation has made the grid less reliant on the bulk transmission grid. Local resource

Electric Power Systems Resiliency
https://doi.org/10.1016/B978-0-323-85536-5.00003-5

availability and nonavailability of the transmission network in certain areas have motivated the community to develop islanded grid to serve a certain rural community, which is also known widely as rural μ-grids. Many communities around the world are openly accepting such an approach. In the event of disastrous events, if the distribution network is resource-rich, people have realized the μ-grids provide the ability to separate from the bulk transmission system ensuring survivability (Panteli et al., 2016); and consequently, μ-grids have been growing voice to be able to be utilized as resiliency mitigation solution (Maloney, 2020). The smaller geographical foot prints of μ-grids allow for continuity of service, while LPHI events affect a larger area. It is trivial that some of the loads can be shed to ensure continued supply for the critical infra-structure even in the resource-poor situation. Consequently, μ-grids represent a well-defined electrical system with local generation resources and associated loads that can, when necessary, operate independently in islanded mode. They may or may not be part of a larger grid, creating the main classification of grid-connected and remote μ-grids.

The condition of resourcefulness within a μ-grid may not always be satisfied, and some of the μ-grids can be resourceful compared to the others. Consequently, multiple μ-grids can be connected via transmission/distribution networks exchanging resources among each other (Chanda & Srivastava, 2016; Xu & Srivastava, 2016), improving robustness. Here, even if numerous power lines or μ-grids themselves get outaged, the rest of the μ-grids will still be able to perform, at least with load shedding, enabling robustness of the grid. μ-grids can isolate themselves if the transmission system is suffering from a cascaded outage (Guo et al., 2017). They can pick-uploads from the unhealthy non-μ-grid area (if μ-grid infrastructure exists) dynamically. All these benefits can be achieved remotely while the event is in progress.

Additionally, power lines are often de-energized prior to the event occurrence to avoid the origination of the secondary source of events originating from the power lines (Abatzoglou et al., 2020).μ-grids can keep the lights on in those areas that could not be served due to proactive de-energization and limited availability of tie lines. μ-grids are essentially useful during postevent restoration of the power system, even if the transmission lines becomes unavailable, ensuring faster recovery (C. Chen et al., 2016). However, as already discussed, all of these can be achieved if the μ-grids are resource-rich, capable of being operated as an island backed by advanced telemetry, modern information processing, and equipped with suitable control devices. Consequently, while the emergence of DER technologies, availability of flexible resources, improved manufacturing are certainly an enabler, the ability to connect and disconnect into the grid comes at a tremendous price. Therefore, the utility of using μ-grids as a resiliency resource is a crucial question, and in this chapter, we have tried to focus on the following questions:

- What are the key features of μ-grids that make it a resiliency resource?
- How is μ-grid resiliency evaluated?
- During the different stages of event progression, how will μ-grid operation change to enable resiliency?
- What are the challenges of utilizing μ-grids as a resiliency resource?

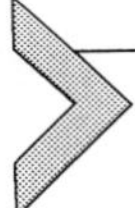

2. Key resources offered by microgrid within distribution system

μ-grids possess a unique set of features that allow for their operation to supplement the system's resiliency. Consider the portion of the distribution system that does not have μ-grid capability and suffers the same outage as the bulk distribution system during an LPHI event. The typical mitigation of LPHI outages is similar to existing distribution system mitigation strategies prescribed by utilities to alleviate grid blackout and improve system stability. Any centralized control that can be executed by the distribution network operator needs to be immediate and constrained to the location of sectionalizers and power sources to avoid cascading outages into the system. Typical operations to reduce discontinuity of service take the form of the following steps:

i. Reconfiguration using switches in the network to serve loads through an alternative source of power. Network reconfiguration changes the topology of the network to connect the disconnected portion of the distribution system to either a healthy circuit with enough capacity.

ii. Use of backup generation to provide service to loads that are disconnected from utility service. These backup generators are usually diesel units that are maintained for intermittent service during emergency conditions.

iii. Crew and resource dispatch to repair infrastructural damage and restore healthy operation of the distribution network. This step is usually very subjective on the nature of the distribution system and associated affected components.

iv. Adaptive islanding allows the distribution network to act as a μ-grid with a small electrical boundary and utilize distributed energy resources to power the μ-grid.
v. Shedding of nonpriority loads alleviates the generation load mismatch and can restore critical loads.

With μ-grid capability implemented, that is, the ability of the distribution network segment to operated isolated from the primary grid using local resources, certain features of the μ-grid can be leveraged to impart resilience. Microgrid architectures are varied, and there does not exist a single approach to utilize all μ-grids as a resilience resource. Instead, the following features can be assessed with location-specific threats applicable to the system and promise to improve resilience.

2.1 Resource flexibility

The inherent design and planning paradigm of μ-grids provide the operational flexibility, increased availability, and the distributed architecture required for the quick restoration and continuation of service to critical loads. The inherent vulnerability of bulk distribution systems also stems from the long distance between the generation source and the loads, contributing to increased susceptibility. The advantage of the operational flexibility is the high power availability of μ-grids during LPHI events where the event's impact is distributed unevenly. Therefore, the boundary of the μ-grid, limited by the switching configurations and location of DERs, can adjust to maintain service. The additional benefit of the smaller footprint of μ-grids also reduces the energy losses commonly observed in bulk distribution systems due to the radial nature and high R/X ratio. The size of the μ-grid envelope is defined by the area covered between the source and generation, ranging from a single building (generator-load pair) μ-grid to a full substation μ-grid with multiple DERs and loads with full integration contribute to the survivability.

μ-grids allow for bidirectional flexibility: upstream flexibility on the resource side and downstream flexibility on the load side. Fig. 7.1 shows the configuration of the μ-grids in relation to the flexible assets. Engineers can leverage the resource flexibility to ensure that the μ-grid can accommodate for resilient operation during LPHI events.

2.2 Energy storage

Energy storage allows the μ-grid to store energy from nondispatchable variable generation in the μ-grids and utilize it as a dispatchable source when

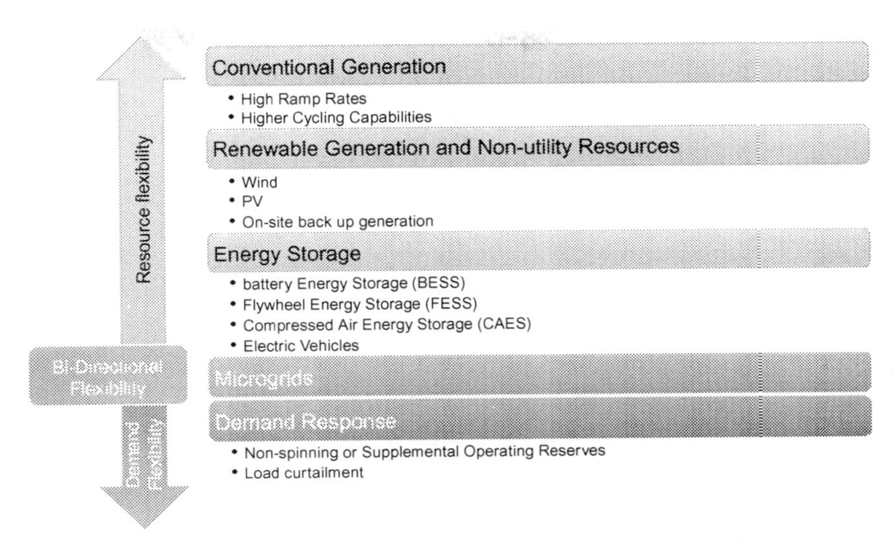

Fig. 7.1 Resource flexibility enabled by μ-grids. *(No Permission Required.)*

necessary. Commonly available in the form of battery energy storage systems, electric vehicles, pumped hydro plants, and more recently electric vehicles, energy storage have become an essential design consideration in μ-grids. Energy storage devices provide a fast response to absorbing the abrupt loss of generation due to LPHI events to reduce generation stress. Also, energy storage systems can provide improved μ-grid stability through voltage and frequency regulation. Typical BESS, implemented in most commercial instances using lithium-ion technology, the major issues in adopting storage into μ-grids seems to be the cost of the batteries, storage capacity, and operating times. Careful planning assessment needs to be considered to utilize energy storage as a resilience improving investment. The planning study should also include determining where in the μ-grid the energy storage unit be placed in the optimal location needs to be resolved as proposed in Kim and Dvorkin (2019).

2.3 Distribution automation

The unpredictability and variability introduced by DERs in μ-grids was once seen to add unwanted complexity to the μ-grid operation. Distribution automation (DA) is a potential solution to mitigate this problem. In addition to introducing control and monitoring to μ-grid assets, DA allows for enabling resilience by leveraging sensor networks, communication networks, controls, and data analytics.

DA as reported in available literature shows the following capabilities:

i. Improved outage management
- Remote fault location and diagnostics
- Automated feeder switching
- Outage status monitoring and notification
- Optimized restoration dispatch

ii. Voltage and reactive power management
- Integrated voltage and volt-ampere reactive (VAR) controls (IVVC)
- Automated voltage regulation
- Conservation voltage reduction (CVR)
- Real-time load balancing
- Automated power factor corrections

iii. Frequency and real power management
iv. Equipment health monitoring
v. Coordination of μ-grid assets

Each of these are potential application of the DA and can be utilized to enable resilience of the μ-grid.

3. Assessment of distribution system resilience with microgrid

μ-grids as a resilience resource can be realized in three stages, each corresponding to the progression of the event in the temporal horizon. In this section, we describe the resilience analysis process of μ-grids before, during, and after the LPHI event. The characteristics of the μ-grids that enable resiliency are:

- **Preparedness:** The property of the system's readiness for the incoming disruptive event. This multidomain property allows for the various mechanisms in the μ-grid to be ready to respond when the event progression starts.
- **Robustness:** The property of systems to resist change in topology when subject to stress
- **Absorption:** The property of the system to resist discontinuity of service when subject to an LPHI event
- **Response:** The property of the system to evaluate and select the appropriate control action to reduce the impact of the LPHI event and improve performance after the LPHI event has passed.
- **Recovery:** The property of the system to regain blue sky performance with minimum time.

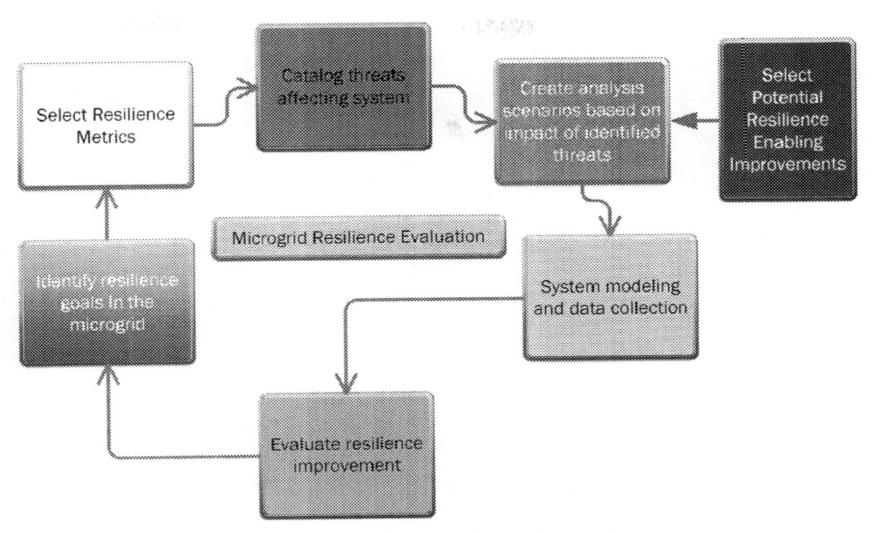

Fig. 7.2 μ-grid resilience evaluation. *(No Permission Required.)*

Careful planning study to assess μ-grid needs to be conducted to justify the initial capital investments in commissioning μ-grid capabilities in a distribution grid. Microgrids to alleviate reliability concerns are designed and implements to reduce the impact of upstream outages in the distribution network and reduce the loss due to the energy not served to the loads. Subsequently, the assessment of μ-grid resilience provides a business case to utilize μ-grids as a resilience resource. This can be quantified through the resilience assessment process shown below in Fig. 7.2. A special case of the resilience evaluation framework is the AWR resilience metrics-Anticipate metric for preevent resilience based on preparedness and ability to anticipate damage to the system; Withstand metric for during event resilience based on robustness and optimal utilization of resources to mitigate LPHI impact and Recover metric for the postevent resilience to address the rapidity and magnitude of the critical load restoration effort as presented in Fig. 7.3.

3.1 Preevent resiliency

The resiliency of the μ-grid before the event is affected explicitly by the ability of the system to anticipate and be prepared for the incoming threat event. At this stage, the μ-grid is expected to work under normal conditions or in "blue sky" conditions. The term blue sky is used to separate between the definitions of reliability and resilience as there can be high resilience and

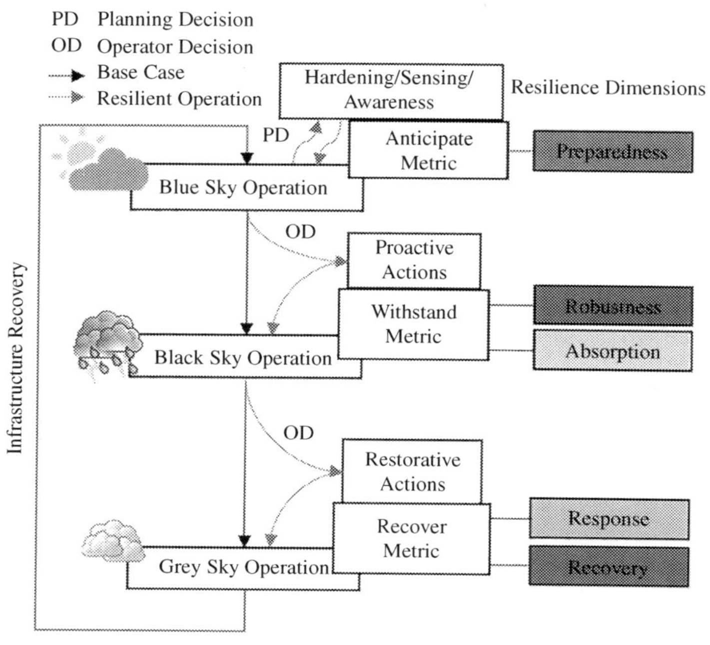

Fig. 7.3 Multitemporal resilience framework. *(No Permission Required.)*

low reliability in blue sky operation and high reliability and low resilience in normal operating conditions. Most of the preparedness strategies to improve μ-grid resilience in the blue sky mode are performed in the form of infrastructure hardening—the process of adding robustness, security, and stability of electrical assets to prevent failures due to the physical effects of the LPHI events. These include "undergrounding" of electrical poles, the elevation of flood-risk assets like generators and switch-gear, vegetation management, and overdesigned construction practices. In addition to hardening, planning efforts such as adding redundancy in the form of adding additional distribution feeders and associated switches, improving resourcefulness through the addition of DERs, better situational awareness through sensors and controls allow for increased resilience in the μ-grid.

Note: The ability of the system to anticipate, be prepared for, and mitigate the impact of the event is quantified by the anticipate metrics.

In order to assess resilience before the event, the system needs to be reviewed and assessed for metrics that will quantify system preparedness and the ability to anticipate, be prepared for, and mitigate the impact of the event. Preevent resiliency analysis helps develop business cases for

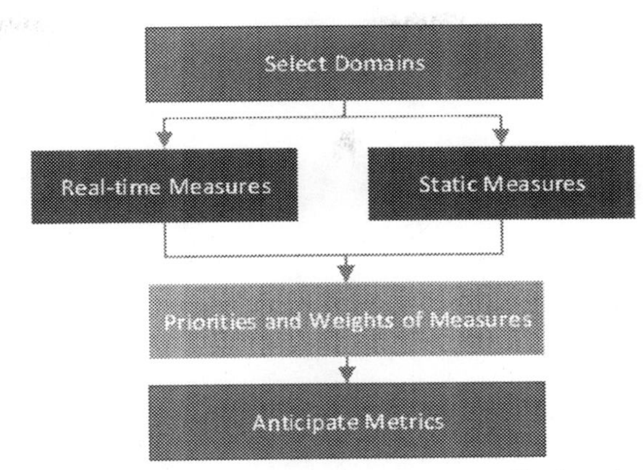

Fig. 7.4 Development of the anticipate resilience metric. *(No Permission Required.)*

resiliency improvement planning studies. The interdisciplinary nature of resilience requires the assessment to include domains that directly affect the μ-grid resilience. These domains include but are not limited to transportation system, fuel supply chain, cyber-communication systems, repair crews, and water distribution systems. These domains have many internal variables that can alter the resiliency of the system but can make the analysis data-intensive and complicated. The use of a composite preevent resiliency metric $\mathfrak{R}_{pre}$ that is objective-based and considers all domains involved is required. This preevent resilience metric should be developed as shown below (refer Fig. 7.4).

Factors from individual domains are selected based on careful consideration from all stakeholders. This process of careful elicitation can be exhaustive and time-consuming due to the system and threat-specific nature of the system in question.

3.2 In event resiliency

A few moments before the event progression starts, the system is operating in blue sky mode and is serving all its critical loads and is meeting some performance standards particular to that μ-grid. The ability of the system, from this point onward, to withstand the impact of the event, thereby ensuring continuous service to critical loads is quantified by the withstand metrics. The first step in the withstand metric evaluation is the selection of relevant resilience indicators that can increase or decrease resilience. The second step is to apply a threat impact analysis to observe the change in resilience indicators before and after the threat. This provides the user with a set of alternatives that can be passed through a multicriteria decision-making tool that can provide a comparative analysis of how the indicators vary between the alternatives. For a typical μ-grid, the resilience indicators pertaining to the withstand metric are:

- **Critical Load Count (CLC):** The number of critical loads still in operation while the event happens
- **Critical Load Rating (CLR):** The size of critical loads still in operation while the event happens
- **Total Available Generation (*G*):** Total generation available for dispatch to critical loads
- **Critical Load Demand (*D*):** Demand of the system
- **Topological Score ($\mathfrak{R}_T$):** A composite score indicating the various graph theoretical measures

A number of graph-theoretic techniques are available for engineers to choose when assessing μ-grid resilience. Let us consider a simple factor analysis using graph-based characteristics of the μ-grid to assess resilience.

3.2.1 Graph theoretic approach to ensure microgrid robustness

For the extraction of topological resilience indicators, the μ-grid is represented as a graph G of N nodes and E edges with the adjacency matrix A. The node elements in the system are transformers, switches, bus elements, usually in the form of switch-gear and circuit selectors. The edge elements are wires connecting the nodes together. Several methods are proposed in current literature to use graph theory techniques to analyze the robustness of the power grid (Motter & Lai, 2002; Solé et al., 2008). In the proposed topological analysis, the network is studied as an undirected and unweighted graph. The weight or the importance of the node are considered in the subsequent evaluations as loads are classified as high priority, medium priority, and low priority loads. The topological resilience indicators selected for this analysis are chosen to indicate how well connected the system is and how much perturbations can the system withstand. These indicators are chosen to represent the size, distribution, node, and link connectivity statistics of the graph.

Graph Diameter (*D*). The maximum eccentricity or the greatest distance from any two vertices indicates the size of the graph.

$$D = \frac{2|E|}{|N|(|N|-1)} \tag{7.1}$$

Degree distribution ($<k>$**).** The degree of a node is the number of nodes connected to it. The connectedness of the graph is indicated by the degree distribution, which is the probability distribution of the degree over the entire network. The average degree distribution indicates the number of feeders arising out of a particular node. A high value of the average degree distribution indicates that there are nodes with high connectivity to other network nodes.

$$\langle k \rangle = \frac{2|E|}{|N|} \tag{7.2}$$

Average betweenness centrality (C_b**).** Betweenness centrality of a node x is defined as the number of shortest paths between all pairs of nodes in the connected graph passing through the given node x. It is a measure of node importance and indicates how many shortest paths are dependent on the nodes present. The average betweenness centrality provides a clue to how susceptible the graph is to perturbations or failures that can possibly sever connections between nodes as it is not favorable to have a node with high betweenness centrality fail in the system.

$$C_b(i) = \sum_{i \neq j \neq k} \frac{\sigma_{jk}(i)}{\sigma_{jk}} \tag{7.3}$$

Percolation threshold (f_c**).** Percolation theory can be employed to study the robustness of the network. Chanda and Srivastava describe the infinite-dimensional percolation analysis of a graph which is subject to random removal of nodes which is denotes by f (Chanda and Srivastava, 2016). This random removal is representative of an unfavorable event. For such study, it is observed that there exists a critical fraction of nodes removed f_c for which the graph degrades into individual isolated clusters. This critical fraction of nodes is called the percolation threshold shown below as an approximation using statistical mechanics approach (Radicchi, 2015).

$$f_c = 1 - \frac{1}{\kappa_0 - 1} \tag{7.4}$$

where $\kappa_0 = \langle\kappa^2\rangle / \langle\kappa\rangle$ and $<\kappa^2>$ is the square of the standard deviation of the degree distribution of the network. The critical fraction of nodes in this graph theoretical analysis should not be confused with critical loads, which are high priority loads that require nondiscontinuity of service during unfavorable events. However, these critical nodes are highly influential in the robustness of the system.

Algebraic connectivity (λ_{alg}). Also called the Fiedler value, it is the second smallest eigenvalue of the Laplacian matrix of the graph. The Laplacian matrix is the sum of the degree matrix D and the negative of the adjacency matrix A. The elements of the Laplacian are given by degree of the node i at diagonal (i,i) positions and by -1 at nondiagonal positions.

Once the topological resilience indicators are extracted, a vector of the indicators is obtained for each scenario that is analyzed. This is represented as

$$\vec{\Re}_{\tau} = [f_c, D, \wedge_2, C_B, \langle \kappa \rangle] \tag{7.5}$$

This vector represents the topological component of the resilience analysis. For each of the system configuration, threat scenarios or study cases, a new vector is created.

3.3 Postevent recovery resilience

The third stage of the event occurrence, when the threat has subsided and the system is trying to recover from the damaged state to the state of normal operation. In this stage, the response and recovery of the system is to be quantified. The recovery metrics depends on the rapidity of restoration, redundancy of resources, and resourcefulness of assets. The postthreat recovery of the system starts with the evaluation of system damage. The survey produces a damage report enumerating the number of damaged assets, including poles, lines, transformers, and switches, and the corresponding location.

With the postevent damage assessment, the recovery of the μ-grid would be a twofold process. The loss of generation would be rectified by the dispatch of the DERs or through blackstart restoration. Then the most resilient μ-grid restoration options need to be evaluated to be selected. As a use-case, let us consider the addition of automated switches that can be used for reconfiguration. With different configurations available, a similar method to the withstand metrics can be employed where the various rapidity and resourcefulness applications can be selected and compared. A typical recovery resilience calculation would be performed as follows. A factor extraction on several resilience indicators needs to be collected, such as the recover cost (RC)—the cost of equipment, labor, and material for the repair of the "downed" assets. The factor CLR is the weighted number of critical loads restored. The repair time RT is based on the type of equipment to be repaired. Since the methodology uses AHP for the computation, the relative time for the time and cost is sufficient to make a decision on the resiliency score. The repair time is the sum of the repair time for the equipment, time for the crew to get from the crew station to the equipment $T_{accessibility}$ and T_{OH}, the overhead for the crew to assemble the repair/recover work. The repair time is assumed to be 1 hour for a feeder, 4 hours for a pad-mounted switch, and 5 hours for a transformer.

$$\vec{\mathfrak{R}}_{\tau} = [RC, RT, CLR, SO, T_{SO}, \mathfrak{R}_{\tau}] \tag{7.6}$$

The number of switching operation required to restore the load (SO) and the time for the switching operations indicated the rapidity of the restoration process. These are computed for available restoration paths for each scenario. The use of real-world repair costs and times is not imperative because the AHP is a comparative process and does not require the factors to be accurate. This provides the Distribution Network Operator (DNO) with operational decision-making assistance on the most resilient restoration scheme.

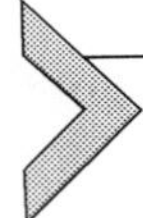

4. Strategies for enabling distribution system resiliency with microgrids

While an LPHI event is on the horizon, the primary objective of the operator is to let the system "bend" proactively (also known as preventive techniques) (Panteli et al., 2017) and adapt to the looming threat to avoid the future cascading failure of the power system (Guo et al., 2017). The μ-grid, in this regard, emancipates a part of the grid to be operable in isolation. The AWR framework discussion clearly divides the entire event temporal horizon vis-à-vis resiliency improvement strategy road-map into three stages. The first stage is called on immediately following the situational awareness signal received from the resilience monitoring system. If the power system resiliency deteriorates, the power system operator will switch from economic operation mode to resiliency mode, where minimization of the load curtailment, maximization of energy served, or minimization of energy not served becomes the main objectives (Bhusal et al., 2020). The loads can also be shredded based on their criticality, resource availability, and ability to form a μ-grid. Operational crews and moving diesel generators (Wood, 2020a, 2020b) are also deployed simultaneously for safety-related de-energization and systemwide reconfiguration, which can also include manually formed μ-grids. It is also notable that the entire distribution network may get isolated from the transmission grid following the dissipation of the disastrous event. Repair crews can be prepositioned (C. Chen et al., 2017) to ensure faster restoration. Nevertheless, this stage is required to be carried out sufficiently in advance for ensuring the in-event safety of the operational crews. The second stage commences with the event striking the critical power infrastructure. The crews' limited availability enforces that the majority of the requisite deployed operation will be carried out remotely.

Once the disaster has precipitated, the repair crews can be deployed to estimate the damage and prioritize the recovery and re-energization process, considering predeployed moving diesel generators and prepositioned crews. Here, restoration of the network necessitates the availability of, as discussed, sufficient black start capability (Schneider et al., 2017) for being able to be operated as μ-grid, or availability of the transmission network. Therefore, depending on the distribution network's operability as a μ-grid, there are two main facets for system operation in this stage (Amin Gholami, Aminifar, & Shahidehpour, 2016). If the network is incapable of operating as a μ-grid, with transmission network outage, a top-down approach needs to be deployed. Here, the network can be restored only after reconfiguration of the transmission network. Restoration of loads would also be carried out after due reconfiguration of the distribution network, only after the primary substation is re-energized. If the network can operate as a μ-grid, restoration of loads within the distribution network can be carried out independent of restoration of the transmission network. The speed of restoration would, of course, be limited by local resource availability. Typical objective function considered by the system operator in this stage will be the minimization of total restoration time and maximum critical infrastructure restoration (Bhusal et al., 2020).

While the discussed three steps are relatively independent in nature, the action plans rely on the measures taken in one of the earlier stages and are shown in Fig. 7.5. In the rest part of this section, we will discuss the challenges of the described three stages and the utility of μ-grid in their mitigation.

4.1 Proactive management and control

As discussed earlier, preevent control and resource allocation begin with a forecasted situational awareness signal. Although the consensus lies in the difficulty of precise evaluation of the origin of the critical weather events and the probability of event severity (Y. Wang et al., 2016), the recent

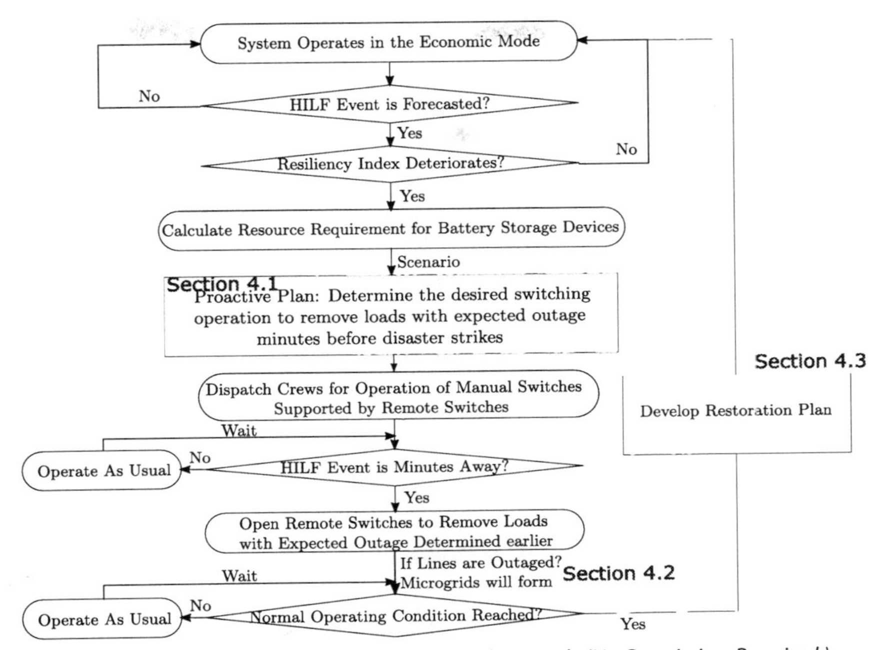

Fig. 7.5 Sequential resiliency management and control. *(No Permission Required.)*

advancement in the accuracy of the weather models has significantly improved the short-term prediction accuracy. In this regard, the use of predicted wild-fires propagation path into an early warning signal for the transmission system outage is already discussed in the literature (Dian et al., 2019), and power utility companies in the United States are actively seeking to develop such early warning signal for the distribution network and use it to improve networkwide resiliency (Boston Consulting Group, 2020). Literature, such as Guikema et al. (2014), Liu et al. (2005), and Nateghi et al. (2014), utilizes statistical models to estimate power outages, which can be utilized to allocate resources, including the deployment of μ-grid to reduce outage duration, based on historical in-event monitoring data (Ji et al., 2016). Even if a disaster strikes energized poles and wires, especially in the case of hurricanes, snowstorms, typhoons, windstorms, floods, and lightning storms, the component failure rate varies with the intensity of hazardous forces. A generic fragility curve can be utilized in this regard for depicting the failure probability of the equipment (Panteli et al., 2016). The intensity of the weather events can also significantly vary spatiotemporally (Y. Wang et al., 2016). Contrarily, the wildfire events directly affect the flow through the transmission lines (Choobineh et al., 2015), and the lines are required to be dynamically rated (Trakas & Hatziargyriou, 2018), forcing the distribution to operate as μ-grid. Multiple μ-grids can be connected to each other, forming a multi-μ-grid (Chanda & Srivastava, 2016; Schneider et al., 2017). Nevertheless, for the event modeling purpose, the propagation of both of these kinds of weather events can be treated as the Markov decision process (Bertsimas et al., 2017; C. Wang et al., 2017).

In the absence of long-term realistic prediction models, robust optimization as a part of proactive management (Gao et al., 2017) is widely used in the literature. As discussed, since the action plans rely on the measures taken in one of the earlier stages, simultaneous consideration of preevent resource allocation (symbolizing "wait and see") with real time in-event dispatch (symbolizing "here and now") can also be considered as a part of robust optimization (A. Gholami, Shekari, et al., 2016; A. Gholami et al., 2019). This enables the predictive-corrective action plan in the decision-making. As discussed in the AWR framework, the availability of the number of lines in a network significantly affects resiliency, and therefore, the removal of all the to-be-affected lines may not be realistic. Consequently, one may also ensure a certain minimum number of vulnerable lines within a distribution grid remain connected in anticipation of an event while ensuring minimal state transition, line outages, and load curtailment (Amirioun et al., 2018).

The discussed proactive crew mobilization (Maryland Energy Administration, 2020) is also driven by safety-related de-energization, expected damage (in terms of the value of the lost load, VOLL), load criticality (loads, such as hospital, water treatment facilities have higher priorities), manual μ-grid formation, the crew dispatching cost, and crew availability. Often, some of the energized equipment is required to be de-energized a priori to avoid origination of secondary disaster (e.g., public safety power shut-off, or PSPS, events in the state of California, United States (Abatzoglou et al., 2020)), or hasten postdisaster recovery (e.g., preevent generator shut down before tsunami and snow avalanche in Alaska, United States). In this case, the reconfiguration is needed to be carried out with sufficient delay to ensure customers are served for the longest possible duration. These operational crews can also facilitate the creation of μ-grid as a defensive islanding strategy, where, even if a part of the grid were required to be isolated, the customers would remain energized during the disaster. This strategy can also help us alleviate cascading outages. The presence of both remote and manually operable switches can be considered in this effort, where operational crews are dispatched insufficient advance so that their safety can be ensured (Arab et al., 2015). Furthermore, the dispatch of the crews and moving diesel generators (Martini, 2014) need to be coordinated. In case the repair crews are dispatched a priori, they must be safely located sufficiently away from the hazard zones (Arab et al., 2015). Although these moving diesel generators can have black start capability, limited fuel availability (Gao et al., 2017), and limited charging capability of batteries (Pandey et al., 2020) are also major concerns for the isolated μ-grid survivability. Additionally, the limited availability of other resources also enforces the available local resources with the μ-grids to be scheduled appropriately to reduce the downtime for the critical loads (Rahman, 2008). Furthermore, if isolated μ-grid operation is looming, the operator might schedule their resources conservatively via resiliency cuts (Khodayar et al., 2012) (by limiting utilization of certain resources above the threshold, precharging batteries (Pandey et al., 2020), etc.).

In the following part of this subsection, we will describe the proactive decision-making strategy through an example. From Fig. 7.5, we observe that the system would continue to operate in the economic model until an LPHI event is forecasted. If the withstand resiliency metric for the concerned system deteriorates, the system will jump to resiliency mode, where the operator aims to supply as much critical load as possible within the system. In this mode, the operator will initially estimate the nodes within the distribution system with the expected outage. If any generator is located within this region, in order to achieve postdisaster expedited recovery, those generators will also be taken out from the grid. Given the finiteness of the available switches, it is expected that some of the additional set of nodes will also be outaged. As a part of operating both manually and remote operable switches, to ensure crew safety, requisite manual switching operations are required to be carried out significantly earlier in the temporal horizon. However, it is also not recommended to disconnect the loads with the expected outage (LwEO) several hours before. This motivates us to solve this problem in two stages. The first stage deals with the operation of manually operable switches, which are assisted by the remotely operable switches, which are to be deployed significantly ahead of the event. Second stage deals with the isolation of LwEO through remote switches, which can be modified as the forecast gets revised. Therefore, this approach utilizes a flavor of predictive- corrective approach. It may so happen that following the disconnection of LwEO, some of the unaffected parts of the network need to be operated as a μ-grid (if designed a priori). The LwEO remains connected to the main grid through remotely operable switches (if that

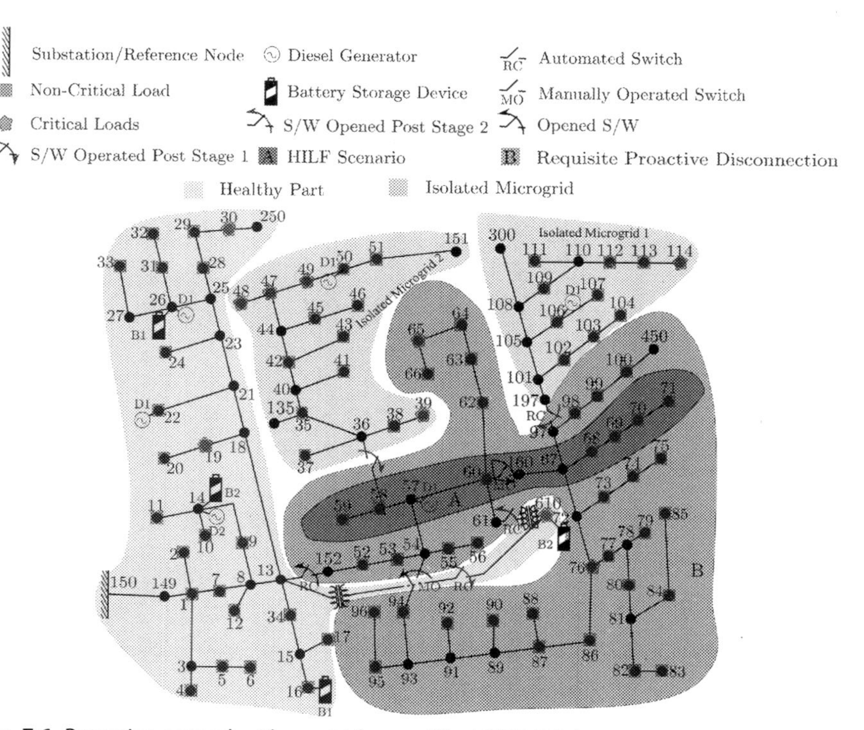

Fig. 7.6 Proactive control with μ-grid for modified IEEE 123-bus system. *(No Permission Required.)*

segment is not designed to operate as a μ-grid) while ensuring the flow through the associated switch is zero. This will ensure that in the event the LwEO is isolated, the operation of the healthy part of the system will not be disturbed.

As shown in Fig. 7.6, the proposed strategy has been depicted using a modified IEEE 123-node system. Locations of manually and remotely operable switches are given. Grayed-out section with the symbol, A, identifies HILF estimated outage scenario. Due to the absence of switches, the entire light grayed region identified with symbol B will be outaged. Highlighted areas in pink color are connected through the LwEO region, and hence will be required to be operated as μ-grids when the event is in progress (if designed). Critical loads connected at node 610 will be connected into the main grid through the associated manually operated switch. To ensure that subsequent disconnection of LeWO, the healthy part will not be disturbed, one needs to ensure that flow through switches 36–58, 97–197,

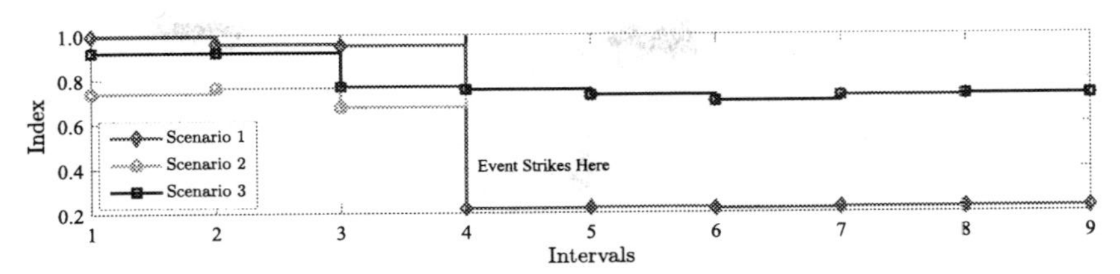

Fig. 7.7 Temporal variation of the withstand metric for modified IEEE 123-bus system. *(No Permission Required.)*

and 61–610 is zero following stage 1, will be opened minutes before disaster strikes. Furthermore, the LwEO will be continued to be supplied in case the HILF event never strikes. Furthermore, since the stage 2 plan is implemented only through remote switches, changes in weather prediction will not be harmful to the operational crews, and, is beneficial due to correction possibility. Therefore, it is clear that the discussed two-stage approach (scenario 3) is beneficial compared to the single-stage fire-and-forget (scenario 2), or no action (scenario 1) approach, and is shown Fig. 7.7.

4.2 Postevent recovery

Although a μ-grid is designed to disconnect from the grid autonomously, it is expected to operate in grid-connected mode during normal operating condition (Z. Wang & Wang, 2015). Once the contingency event strikes the distribution network, a part of the grid can become de-energized, as a part of an antiislanding protection scheme, or operate as a μ-grid. Availability of system status information, infrastructure resources (reclosers, energy storage system, DERs, mobile generators) facilitates the seamless formation, and successful operation of such μ-grid (C. Chen et al., 2017). In this regard, Gouveia et al. (2013) utilize emergency demand response from responsive loads (including EVs) for successful primary frequency control within a μ-grid establishing continuous load generation balance. The use of CHPs, and diesel generators for postevent operation is also a widely accepted choice (KEMA, 2014; Maloney, 2020). Depending upon the R/X ratio of the distribution network, $P-V$ and $Q-\delta$ droop characterized can be utilized (Abdelaziz et al., 2014) for power balance while ensuring protection devices does not get triggered.

Here, the grid can also be equipped with self-healing technologies, such that once a de-energization event has occurred, the μ-grid can intelligently pick-up loads within its vicinity (assuring survivability) (Adibi & Fink, 2006). The resiliency could be further improved by dynamic creation of μ-grids, forming networked μ-grid, multienergy μ-grids, or energy hub (Hussain et al., 2019). The dynamic μ-grid creation is especially favored as a reactive mechanism and is often facilitated by software algorithms for controlling the switches (Simonov, 2014). Artificial intelligence, multiagent (Dehghanpour et al., 2017), fuzzy logic (Hussain et al., 2017)-based techniques can be utilized to facilitate demand response, DER, and storage system dispatch as a part of advanced operation strategy within a μ-grid. In case multiple μ-grids are formed, multiple μ-grids may self-organize into a cluster (Ding et al., 2017; He & Giesselmann, 2015) for optimal power-sharing (Zadsar et al., 2017), if the capability exists. AC, DC, and hybrid-AC-DC μ-grids can be utilized for resiliency improvement (Pannala et al., 2020). Such a power sharing approach ensures the self-sufficiency and stability of the network as a whole.

The distribution network in its entirety can be decomposed into multiple μ-grids (C. Yuan, 2016). However, as discussed earlier, limited fuel availability limits the operating modes of these μ-grids and can be categorized into basic autonomous, fully autonomous, and networked μ-grid (Zia et al., 2018). While the disaster is in progress, the μ-grid might limit its resource provision to critical facilities. The use of the distributed multiagent system to achieve such a self-healing ability for a μ-grid has been discussed in Colson et al. (2011). Given that the μ-grids remain energized during the progressing event, it can serve black-start capability to re-energize disrupted main generators (Schneider et al., 2017). Therefore, depending upon its capability of the resource within a μ-grid, and deployed smart switches, and tie-switches, it can be categorized into three major categories (Schneider et al., 2017): (i) local resource, (ii) community resource, (iii) black-start resource. In case some of the switches are not automated, the system operators can also facilitate μ-grid formation remotely. Once the emergency transpires, it will switch back to the normal mode. Reliance on the communication network in such a venture can be facilitated by adopting a distributed approach (C. Chen et al., 2016). The utilization of centralized communication architecture may not be suitable in this regard due to network-level vulnerabilities.

4.3 Restoration process

In conjunction with the traditional distribution service restoration as a part of the outage management system (OMS) (Singh et al., 2017), here the post contingency recovery begins with damage assessment through ground check, aerial survey, and customer trouble calls, SCADA (protective relays, fault indicators, etc.), AMI, μ-PMU, social media, and advanced resiliency monitoring system (ARMS) as a part of the advanced distribution management system (ADMS). All of this information can be fused to provide grid status as a part of the decision support tool. Large scale deployment of monitoring devices in a smart grid facilitates such treatment (Mohamed et al., 2019). Following this, the operator calculates the optimal route for crew dispatch, crew schedule to mobilize crew from one site to another while minimizing the repair and restoration time for the critical loads. Crew and resource budget is also a major concern in this step and often responsible for delayed recovery (W. Yuan et al., 2016). Contrary to the preevent proactive crew dispatch (if carried out), the state of system outage is already deterministic. Furthermore, one also needs to ensure that antiislanding protection for each of the backup generators/DERs is present, such that the safety of the repair crew is never compromised.

In the absence of local DERs, to speed up the restoration process, moving diesel generators are often deployed (Lin et al., 2019). Additionally, deployment of generators needs to be well-coordinated with switching strategy (B. Chen et al., 2019) to enable the creation of multi-μ-grid, which was not possible due to the absence of sufficient grid intelligence and generators with black-start capability. Such multi-μ-grid framework can significantly deviate from the traditional sequential predetermined restoration priority list (Freeman et al., 2010) and move to parallel service restoration, hastening the entire process. Note that, in this case, crew routing vis-à-vis vehicle routing problem, diesel generator dispatch, and power system re-energization need to be carried out simultaneously, similar to a multienergy system, which has gained significant attention in recent years.

It is also notable that alongside the destruction of electricity and communication infrastructure, other unity infrastructure can breakdown as well (Girgin & Krausmann, 2014). In this scenario, even if gas-fired generators remain unscathed, the generator will be outaged, crippling the capability of a μ-grid (Zerriffi et al., 2007). To circumvent this problem, literature considers information sharing among multiple infrastructure resources (Capozzo et al., 2017). Additionally, one also needs to account for the interdependence among these critical infrastructures. For example, suppose gas pump plants are not treated as a critical facility and restored earlier. In that case, the gas-fired generators cannot be brought online, degrading the power system resiliency significantly (Lin et al., 2019). The required integrated approach for designing a restoration plan for the grid should also adapt to the changing disaster landscape. In case of proactive decision-making, this integrated solution will be part of the "wait and see" problem.

5. Barriers and challenges

μ-grids, enabled by the advanced decision support system and availability of copious local resources, can supplement the primary grid unavailability in the wake of a disaster, leading us to enhanced emergency preparedness of the grid (Xu & Srivastava, 2016). Isolation from the primary grid can help protect the load center from the possible cascaded outage, even in the wake of a cyber-attack. Along with the utilization of physical properties of the power system, μ-grid control center (MGCC) facilitated by decision support tool from local measurements can also utilize advanced software algorithms to manage resources within its premises, dynamically pick up load through increasing its boundary (Hussain et al., 2019), and create a μ-grid cluster, collaborating in a power-sharing approach. The use of clean energy resources during normal operation also facilitate GHG emissions reduction. Always-on power can ensure (Maloney, 2020): (i) continued access to food and water, (ii) operability of clinics, pharmacies, hospitals, (iii) long-term availability of transportation fuel, (iv) continued availability of distribution retail supply-chain, (v) last-minute provision of necessary home-safety and repair tools. Furthermore, many companies, such as Schneider Electric, ABB, Siemens, General Electric, Alstom, Tesla, and Google as of date are developing and deploying their prototype. In the aftermath of the California wildfire and PSPS events, μ-grids have gained a lot of attention from the California Public Utilities Commission (Wood, 2020a, 2020b). However, reaping such huge benefits comes with an enormous price. This section will focus on various barriers, concerns on economic viability, policy requirements, and regulatory barriers, specific to μ-grids as a resiliency resource.

5.1 Barriers

Several implementation challenges require special consideration to achieve the desired level of resilience during major events. These barriers are majorly divided into five categories and are discussed below:

5.1.1 Renewable uncertainty

While higher penetration of renewable-interfaced DERs, coupled with improved predictability of renewable energy resources, enhances the distribution network's self-sufficiency, renewable uncertainty is still a concern during natural disasters. Since event origination and dissipation time cannot be predicted accurately, compensation for the renewable induced variability becomes hard to manage, and the conservative operating solution becomes very expensive. Additionally, during certain weather events, the renewable output becomes zero, leading to resource deficiency and in-event infrastructure damage, especially to renewable generators, blockage of renewable generators due to uprooted trees would lead us to rely on a new set of resources in the postevent aftermath. Furthermore, traditional scheduling frameworks (including stochastic model) consider longer scheduling horizon (Hussain et al., 2019). In-event higher level of uncertainty associated with renewable resources makes real-world deployment of these solutions difficult. Therefore, robust uncertainty handing and predictive-corrective control approach need to be incorporated in the modeling to let the system operate at the desired level of resiliency independent of its operating condition.

5.1.2 Control issues

While it is almost universally accepted that μ-grids can operate as an islanded grid, can connect and disconnect from the main grid, it is a challenge to determine when do we allow the μ-grid to unilaterally disconnect (as a part of "intentional islanding") and when do we allow it connect back to the main grid. If the action is reactive, a temporary outage or voltage dip might be observed within the μ-grid, which will clear itself following the ramping up of local generators. In the case of insufficient responsiveness, the μ-grid would have to utilize its local black start capability, and the internal monitoring system and protective relays, which can often be cost-prohibitive (Hussain et al., 2019). The presence of numerous local DERs makes internal coordination to become challenging, and a comprehensive understanding of the interconnectedness of all the components becomes essential. Furthermore, the majority of DERs are inverter-interfaced, and after disconnection from the primary grid, it reduces its inertia significantly. Consequently, in the event of insufficient flexible fast ramping resource, even if load-generation balance is ensured, the ROCOF of the system will be very high, and traditional protection system settings would also be required to be updated accordingly. This is also true when a μ-grid is dynamically expanding its territory. Furthermore, a fixed μ-grid boundary may not be suitable in a dynamic environment.

5.1.3 Communication infrastructure

Coordination among a multitude of local DERs, requisite local measurements as a part of decision support tool, crew-coordination, as a part of MGCC requires communication links with real-time control capability. Depending on the level of sophistication, one can select the best option among numerous possible options such as WiFi, Bluetooth, Zigbee, passive optical network, and mobile communication technologies (Zia et al., 2018). Various communication architectures, such as centralized, decentralized, hierarchical, and distributed methods, can be utilized. While centralized frameworks are advocated for their ability to better manage local resources (Khodaei, 2014) within a μ-grid, it is well known that lack of redundancy makes centralized framework especially susceptible to communication failure, which would be the case especially in the wake of natural disaster or cyber-threats. While decentralized structure can help us alleviate some of the challenges, reduced visibility reduces its cost-effectiveness (Hussain et al., 2019). Software-defined networking is also proposed in the literature for enhanced resiliency (Jin et al., 2017). Distributed coordination schemes are known to be resilient to communication link failure and have also been considered in the literature. Therefore, there is a growing need to develop robust and event-agnostic communication infrastructure for the successful deployment of μ-grids as a resiliency resource.

5.1.4 Computational need

The increasing complexity of the problem requiring efficient dispatch of multiple renewable interfaced DERs, transmission line switching, crew-dispatch coordination requires sophisticated software algorithms for management. The underlying algorithms need to be robust and need to account for the simplification utilized for disaster planning. The developed decision from the software algorithms needs to be further verified utilizing high-fidelity models before deployment to satisfy safety-related concerns. While distributed optimization techniques can be utilized in this regard (Y. Wang et al., 2016), associated techniques for power system resiliency are still in their infancy. The plans need to be constantly updated with changing operating scenarios. The model also needs to account for the possibility of data corruption due to the possibility of a communication outage.

5.1.5 The need to aggregate: Cost-benefit analysis

Over the years, the critical facilities have been using backup diesel generators (Maloney, 2020). However, one major drawback in this regard would be their high maintenance cost, failing which they become unreliable. In addition to that, during disasters, diesel supply chains can be crippled, leading to their unavailability when needed. While natural gas-based generators are relatively unaffected by the ground-based supply chain, as discussed, those generators can suffer outages due to the breakdown of gas transportation infrastructure (Girgin & Krausmann, 2014). While numerousness of available local resources solves the problem of a single backup generator, and early-adopters have significantly reduced the deployment cost of μ-grids, such a solution is still very expensive due to the utilization of multitude of switches, controllers and communication infrastructure. Nevertheless, multiple retailers can participate in a symbiotic fashion to avail the economy of scale, and such an aggregation model is not cost-prohibitive for small retailers, leading us to μ-grids-as-a-service (MaaS) model (Maloney, 2020). Here, the retailers do not directly bear the cost of μ-grids; rather, it is borne by the developer. This facilitates the retailer to advertise their electricity supply to be superior to the others. The third-party investors or the developers also earn a return on their investment by selling power directly into the wholesale market, when the isolated μ-grid service is no longer needed by the retailer. While a single small μ-grid may be too small to participate in the bulk power market, the aggregation approach with multiple operators facilitates that. Therefore, although disaster-related power outages have become common in recent year, investment into μ-grids are still dependent on carefully looking into possible revenue stream and deployment cost and following conduction of cost-benefit analysis.

5.2 Business models, regulatory barriers, and policy requirements

Higher cost has been demotivating factors for the procurement of μ-grids by the retailers, and consequently, multiparticipation MaaS or reliability-as-a-service (RaaS) models have flourished (Metelitsa, 2018). While μ-grids have proven their efficacy in improving the resiliency of the grid during multiple hurricanes and wildfires in the United States (Maloney, 2020), only fewer cities around the world are leading their way to enable resiliency in their power grid (York & Jarrah, 2020). To the developers, the current rate structure is so marginal that a small variation in price can potentially negate the economic benefit from μ-grid deployment (Borghese et al., 2017). Nevertheless, the European Union, the United States, Australia, and Japan are a few countries worldwide leading in terms of μ-grid-related policies. Especially, Japan, following the Fukushima disaster, has taken an early lead in improving their power grid's resiliency through μ-grid deployment (Marnay et al., 2015).

As reported in the literature, there exist a multitude of business models that can be adopted for the development and operation of μ-grids, and each of the business models has its own set of challenge (KEMA, 2014). Consortium for Electric Reliability Technology Solutions (CERTS) demonstration project has been one of the earliest demonstration projects for utility-owned μ-grid project. However, in many cases, utility companies are not allowed to own generation assets as a part of deregulation, limiting its applicability. Customers owning CHPs, or small generating facilities can be allowed to develop μ-grid, but is often cost-prohibitive due to increased investment in monitoring and protection. One of the recent models has been a multiuser model, where the possible challenge could be franchise right. The hybrid model involves shared responsibility, where the μ-grid authority owns generation, but the loads can be managed by the distribution utility. Under this paradigm, challenges would arise in managing resources when multiple participants are competing for them during the resource-poor condition. Nevertheless, identifying the value stream, contribution partners, ownership and revenue sharing, and overall control has been the major factor for determining the suitable business models for μ-grid (Asmus & Lawrence, 2016).

While newer business models are appearing, increasing the affordability of the μ-grids (Asmus & Lawrence, 2016), and countries are updating their μ-grid related policies (Ali et al., 2017; Wood, 2020b), the regulation is not uniform across the world. As discussed, this has been especially challenging with the multiparticipation framework, existing interconnection process and requirements, existing rate structures, and utility involvement in managing customer demand (Hoffman & Carmichael, 2020; KEMA, 2014). As an example, in case a distribution company has a fixed service territory, and no other entity can serve the power within such predefined territory, the μ-grid operator has to accept the rate structure provided by the distribution company, which can marginalize the profitability of the μ-grid operator. Additionally, in a certain service territory, if the μ-grids generates power for sale, the μ-grid operator might be treated as a public utility and subject to significant regulation, discouraging them from embarking into a μ-grid related project. While the MaaS structure allows multiple μ-grid to remain connected during the emergency condition, and day-to-day management will be carried out by the distribution utility, improving quality proposition for the retailers (Maloney, 2020), a significant reduction in the value-stream will also be discouraging for the μ-grid developers. Furthermore, such a multiplayer structure does not fit well if the utilities are vertically integrated with centralized resources, however, they fit well in the deregulated environment (Hoffman & Carmichael, 2020). Since μ-grids contains a fantastic proposition through resiliency improvement, some utilities may not be supportive of these ventures.

While the progress has been very slow, regulators around the world are legalizing μ-grid development. As an example, California Public Utilities Commission (CPUC) has published their standards, protocols, guidelines, methods, rates, and tariffs that serve to support and reduce barriers to microgrid deployment to emphasize immediate action plans for the utilities to develop μ-grids (John, 2020).

6. Summary

μ-grids provide a unique opportunity for resource-constrained distribution networks to adopt key strategies of resilience by ensuring survivability of critical loads, dampening of impacts of LPHI events, and improving system robustness. With the increased proliferation of distributed energy resources, the formation of self-sustainable μ-grids allows for proactive management of resources to reduce downtime of critical loads through reconfiguration, scheduling of DERs and energy storage, and shedding faulty portions of the system. Even in systems where the impact and duration of LPHI events are not known in advance, feasible islanding and formation of μ-grids take advantage of the uneven distribution of damage in the system to reduce critical loads lost. Utilizing energy storage, distribution automation, data analytics, and situational awareness, μ-grids can employ operational strategies in all stages of the event progression—proactive scheduling and control before the event, resilient mode of operation during the event, and corrective reconfiguration after the event. This book chapter provides the multistage resilience framework for the evaluation of resilience metrics for each stage. The argument for μ-grids as a resilience resource is reinforced through the elaboration of resilience-enabling strategies. The barriers in implementing μ-grids, include but are not restricted to the uncertainty of renewable DERs, communication infrastructure, cybersecurity cost of implementation, and regulatory roadblocks that every utility needs to elicit with stakeholders and customers for the correct action plan. In addition to improving the resilience behavior of the system, μ-grids also provide improved economic operation, socio-economic benefit, improved reliability, and reduced system loss. The operational configuration of μ-grids is varied, and careful consideration of the different options for implementing μ-grids to achieve resilience can be assessed through the μ-grid resilience evaluation framework presented in this chapter.

References

Abatzoglou, J. T., Smith, C. M., Swain, D. L., Ptak, T., & Kolden, C. A. (2020). Population exposure to preemptive de-energization aimed at averting wildfires in northern California. *Environmental Research Letters, 15*(9).

Abdelaziz, M. M. A., Farag, H. E., & El-Saadany, E. F. (2014). Optimum droop parameter settings of islanded microgrids with renewable energy resources. *IEEE Transactions on Sustainable Energy, 5*(2), 434–445. https://doi.org/10.1109/TSTE.2013.2293201.

Adibi, M. M., & Fink, L. H. (2006). Overcoming restoration challenges associated with major power system disturbances—Restoration from cascading failures. *IEEE Power and Energy Magazine, 4*(5), 68–77. https://doi.org/10.1109/MPAE.2006.1687819.

Ali, A., Li, W., Hussain, R., He, X., Williams, B., & Memon, A. (2017). Overview of current microgrid policies, incentives and barriers in the European Union, United States and China. *Sustainability, 9*(7), 1146. https://doi.org/10.3390/su9071146.

Amirioun, M. H., Aminifar, F., & Lesani, H. (2018). Resilience-oriented proactive management of microgrids against windstorms. *IEEE Transactions on Power Systems, 33*(4), 4275–4284. https://doi.org/10.1109/TPWRS.2017.2765600.

Arab, A., Khodaei, A., Han, Z., & Khator, S. K. (2015). Proactive recovery of electric power assets for resiliency enhancement. *IEEE Access, 3*, 99–109. https://doi.org/10.1109/ACCESS.2015.2404215.

Asmus, P., & Lawrence, M. (2016). *Emerging microgrid business models*. Navigant Research.

Bertsimas, D., Griffith, J. D., Gupta, V., Kochenderfer, M. J., & Mišić, V. V. (2017). A comparison of Monte Carlo tree search and rolling horizon optimization for large-scale dynamic resource allocation problems. *European Journal of Operational Research, 263*(2), 664–678. https://doi.org/10.1016/j.ejor.2017.05.032.

Bhusal, N., Abdelmalak, M., Kamruzzaman, M., & Benidris, M. (2020). Power system resilience: Current practices, challenges, and future directions. *IEEE Access, 8*, 18064–18086. https://doi.org/10.1109/ACCESS.2020.2968586.

Borghese, F., Cunic, K., & Barton, P. (2017). *Microgrid business models and value chains*.

Boston Consulting Group. (2020). *Sentient Energy's grid analytics system moves the needle on wildfire mitigation & emergency response initiatives*. Sentient Energy. https://images.assetscem.endeavorb2b.com/Web/PentonCEM/%7B71445f81-29e4-4d97-bb0d-a4b17858353a%7D_Sentient_Energy_Wildfire_Public_Safety_Solution.pdf.

Capozzo, M., Rizzi, A., Cimellaro, I. P., Barbosa, A., & Cox, D. (2017). Earthquake and tsunami resiliency assessment for a coastal community in the Pacific Northwest, USA. In *Structures congress 2017: Business, professional practice, education, research, and disaster management—Selected papers from the structures congress 2017* (pp. 122–133). American Society of Civil Engineers (ASCE). https://doi.org/10.1061/9780784480427.011.

Chanda, S., & Srivastava, A. K. (2016). Defining and enabling resiliency of electric distribution systems with multiple microgrids. *IEEE Transactions on Smart Grid,* 7(6), 2859–2868. https://doi.org/10.1109/TSG.2016.2561303.

Chen, C., Wang, J., Qiu, F., & Zhao, D. (2016). Resilient distribution system by microgrids formation after natural disasters. *IEEE Transactions on Smart Grid,* 7(2), 958–966. https://doi.org/10.1109/TSG.2015.2429653.

Chen, C., Wang, J., & Ton, D. (2017). Modernizing distribution system restoration to achieve grid resiliency against extreme weather events: An integrated solution. *Proceedings of the IEEE, 105*(7), 1267–1288. https://doi.org/10.1109/JPROC.2017.2684780.

Chen, B., Ye, Z., Chen, C., Wang, J., Ding, T., & Bie, Z. (2019). Toward a synthetic model for distribution system restoration and crew dispatch. *IEEE Transactions on Power Systems*, *34*(3), 2228–2239. https://doi.org/10.1109/TPWRS.2018.2885763.
Choobineh, M., Ansari, B., & Mohagheghi, S. (2015). Vulnerability assessment of the power grid against progressing wildfires. *Fire Safety Journal*, *73*, 20–28. https://doi.org/10.1016/j.firesaf.2015.02.006.
Colson, C. M., Nehrir, M. H., & Gunderson, R. W. (2011). Distributed multi-agent microgrids: A decentralized approach to resilient power system self-healing. In *Proceedings—ISRCS 2011: 4th international symposium on resilient control systems* (pp. 83–88). https://doi.org/10.1109/ISRCS.2011.6016094.
Dehghanpour, K., Colson, C., & Nehrir, H. (2017). A survey on smart agent-based microgrids for resilient/self-healing grids. *Energies*, *10*(5). https://doi.org/10.3390/en10050620.
Dian, S., Cheng, P., Ye, Q., Wu, J., Luo, R., Wang, C., Hui, D., Zhou, N., Zou, D., Yu, Q., & Gong, X. (2019). Integrating wildfires propagation prediction into early warning of electrical transmission line outages. *IEEE Access*, 7, 27586–27603. https://doi.org/10.1109/ACCESS.2019.2894141.
Ding, T., Lin, Y., Bie, Z., & Chen, C. (2017). A resilient microgrid formation strategy for load restoration considering master-slave distributed generators and topology reconfiguration. *Applied Energy*, *199*, 205–216. https://doi.org/10.1016/j.apenergy.2017.05.012.
Freeman, L., Stano, G., & Gordon, M. (2010). Best practices for storm response on us distribution systems. In *Proc. 2010 DistribuTech*.
Gao, H., Chen, Y., Mei, S., Huang, S., & Xu, Y. (2017). Resilience-oriented pre-hurricane resource allocation in distribution systems considering electric buses. *Proceedings of the IEEE*, *105*(7), 1214–1233. https://doi.org/10.1109/JPROC.2017.2666548.
Gholami, A., Aminifar, F., & Shahidehpour, M. (2016). Front lines against the darkness: Enhancing the resilience of the electricity grid through microgrid facilities. *IEEE Electrification Magazine*, 18–24. https://doi.org/10.1109/MELE.2015.2509879.
Gholami, A., Shekari, T., Aminifar, F., & Shahidehpour, M. (2016). Microgrid scheduling with uncertainty: The quest for resilience. *IEEE Transactions on Smart Grid*, 7(6), 2849–2858. https://doi.org/10.1109/TSG.2016.2598802.
Gholami, A., Shekari, T., & Grijalva, S. (2019). Proactive management of microgrids for resiliency enhancement: An adaptive robust approach. *IEEE Transactions on Sustainable Energy*, *10*(1), 470–480. https://doi.org/10.1109/TSTE.2017.2740433.
Girgin, S., & Krausmann, E. (2014). *Analysis of pipeline accidents induced by natural hazards*. Eur. Union.
Gouveia, C., Moreira, J., Moreira, C. L., & Pecas Lopes, J. A. (2013). Coordinating storage and demand response for microgrid emergency operation. *IEEE Transactions on Smart Grid*, *4*(4), 1898–1908. https://doi.org/10.1109/TSG.2013.2257895.
Guikema, S. D., Nateghi, R., Quiring, S. M., Staid, A., Reilly, A. C., & Gao, M. (2014). Predicting hurricane power outages to support storm response planning. *IEEE Access*, *2*, 1364–1373. https://doi.org/10.1109/ACCESS.2014.2365716.
Guo, H., Zheng, C., Iu, H. H. C., & Fernando, T. (2017). A critical review of cascading failure analysis and modeling of power system. *Renewable and Sustainable Energy Reviews*, *80*, 9–22. https://doi.org/10.1016/j.rser.2017.05.206.
He, M., & Giesselmann, M. (2015). Reliability-constrained self-organization and energy management towards a resilient microgrid cluster. In *2015 IEEE power and energy society innovative smart grid technologies conference, ISGT 2015*Institute of Electrical and Electronics Engineers Inc. https://doi.org/10.1109/ISGT.2015.7131804.
Hoffman, S., & Carmichael, C. (2020). *Six barriers to community microgrids and potential ways developers can surmount them*.

Hussain, A., Bui, V. H., & Kim, H. M. (2017). Fuzzy logic-based operation of battery energy storage systems (BESSs) for enhancing the resiliency of hybrid microgrids. *Energies, 10* (3). https://doi.org/10.3390/en10030271.

Hussain, A., Bui, V. H., & Kim, H. M. (2019). Microgrids as a resilience resource and strategies used by microgrids for enhancing resilience. *Applied Energy, 240*, 56–72. https://doi.org/10.1016/j.apenergy.2019.02.055.

Ji, C., Wei, Y., Mei, H., Calzada, J., Carey, M., Church, S., Hayes, T., Nugent, B., Stella, G., Wallace, M., White, J., & Wilcox, R. (2016). Large-scale data analysis of power grid resilience across multiple US service regions. *Nature Energy, 1*. https://doi.org/10.1038/nenergy.2016.52.

Jin, D., Li, Z., Hannon, C., Chen, C., Wang, J., Shahidehpour, M., & Lee, C. W. (2017). Toward a cyber resilient and secure microgrid using software-defined networking. *IEEE Transactions on Smart Grid, 8*(5), 2494–2504. https://doi.org/10.1109/TSG.2017.2703911.

John, J. S. (2020). *California's microgrid plan exposes conflicts between utilities and customers*.

KEMA, M. (2014). *Benefits, models, barriers and suggested policy initiatives for the commonwealth of Massachusetts*.

Khodaei, A. (2014). Resiliency-oriented microgrid optimal scheduling. *IEEE Transactions on Smart Grid, 5*(4), 1584–1591. https://doi.org/10.1109/TSG.2014.2311465.

Khodayar, M. E., Barati, M., & Shahidehpour, M. (2012). Integration of high reliability distribution system in microgrid operation. *IEEE Transactions on Smart Grid, 3*(4), 1997–2006. https://doi.org/10.1109/TSG.2012.2213348.

Kim, J., & Dvorkin, Y. (2019). Enhancing distribution system resilience with mobile energy storage and microgrids. *IEEE Transactions on Smart Grid*, 4996–5006. https://doi.org/10.1109/tsg.2018.2872521.

Lin, Y., Chen, B., Wang, J., & Bie, Z. (2019). A combined repair crew dispatch problem for resilient electric and natural gas system considering reconfiguration and DG islanding. *IEEE Transactions on Power Systems, 34*(4), 2755–2767. https://doi.org/10.1109/TPWRS.2019.2895198.

Liu, H., Davidson, R. A., Rosowsky, D. V., & Stedinger, J. R. (2005). Negative binomial regression of electric power outages in hurricanes. *Journal of Infrastructure Systems, 11*(4), 258–267. https://doi.org/10.1061/(ASCE)1076-0342(2005)11:4(258).

Maloney, P. (2020). *Microgrids for the retail sector*.

Marnay, C., Aki, H., Hirose, K., Kwasinski, A., Ogura, S., & Shinji, T. (2015). Japan's pivot to resilience: How two microgrids fared after the 2011 earthquake. *IEEE Power and Energy Magazine, 13*(3), 44–57. https://doi.org/10.1109/MPE.2015.2397333.

Martini, P. D. (2014). *More than smart: A framework to make the distribution grid more open, efficient and resilient*.

Maryland Energy Administration. (2020). *Resiliency through microgrids task force established*. https://images.assetscem.endeavorb2b.com/Web/PentonCEM/%7B71445f81-29e4-4d97-bb0d-a4b17858353a%7D_Sentient_Energy_Wildfire_Public_Safety_Solution.pdf.

Metelitsa, C. (2018). *New business models gain strength with renewed interest in microgrids*. Green 905 Tech Media.

Mohamed, M. A., Chen, T., Su, W., & Jin, T. (2019). Proactive resilience of power systems against natural disasters: A literature review. *IEEE Access, 7*, 163778–163795. https://doi.org/10.1109/ACCESS.2019.2952362.

Motter, A. E., & Lai, Y. C. (2002). Cascade-based attacks on complex networks. *Physical Review E—Statistical Physics, Plasmas, Fluids, and Related Interdisciplinary Topics, 66*(6), 4. https://doi.org/10.1103/PhysRevE.66.065102.

Nateghi, R., Guikema, S. D., & Quiring, S. M. (2014). Forecasting hurricane-induced power outage durations. *Natural Hazards, 74*(3), 1795–1811. https://doi.org/10.1007/s11069-014-1270-9.

Pandey, S., Srivastava, S., Kandaperumal, G., Srivastava, A. K., Mohanpurkar, M. U., & Hovsapian, R. (2020). Optimal operation for resilient and economic modes in an Islanded Alaskan grid. In *IEEE power and energy society general meeting*IEEE Computer Society. https://doi.org/10.1109/PESGM41954.2020.9281707 (Vols. 2020-).

Pannala, S., Patari, N., Srivastava, A. K., & Padhy, N. P. (2020). Effective control and management scheme for isolated and grid connected DC microgrid. *IEEE Transactions on Industry Applications*, *56*(6), 6767–6780. https://doi.org/10.1109/TIA.2020.3015819.

Panteli, M., Trakas, D. N., Mancarella, P., & Hatziargyriou, N. D. (2016). Boosting the power grid resilience to extreme weather events using defensive islanding. *IEEE Transactions on Smart Grid*, 7(6), 2913–2922. https://doi.org/10.1109/TSG.2016.2535228.

Panteli, M., Trakas, D. N., Mancarella, P., & Hatziargyriou, N. D. (2017). Power systems resilience assessment: Hardening and smart operational enhancement strategies. *Proceedings of the IEEE*, *105*(7), 1202–1213. https://doi.org/10.1109/JPROC.2017.2691357.

Radicchi, F. (2015). Predicting percolation thresholds in networks. *Physical Review E*, *91* (1). https://doi.org/10.1103/physreve.91.010801.

Rahman, S. (2008). Framework for a resilient and environment-friendly microgrid with demand-side participation. In *2008 IEEE power and energy society general meeting—Conversion and delivery of electrical energy in the 21st century* (p. 1). https://doi.org/10.1109/PES.2008.4596108.

Schneider, K. P., Tuffner, F. K., Elizondo, M. A., Liu, C. C., Xu, Y., & Ton, D. (2017). Evaluating the feasibility to use microgrids as a resiliency resource. *IEEE Transactions on Smart Grid*, *8*(2), 687–696. https://doi.org/10.1109/TSG.2015.2494867.

Simonov, M. (2014). Dynamic partitioning of DC microgrid in resilient clusters using event-driven approach. *IEEE Transactions on Smart Grid*, *5*(5), 2618–2625. https://doi.org/10.1109/TSG.2014.2302992.

Singh, R., Reilly, J. T., & Wang, J. (2017). *Foundational report series: Advanced distribution management systems for grid modernization, implementation strategy for a distribution management system.*

Smith, A. B., & Katz, R. W. (2013). US billion-dollar weather and climate disasters: Data sources, trends, accuracy and biases. *Natural Hazards*, *67*(2), 387–410. https://doi.org/10.1007/s11069-013-0566-5.

Solé, R. V., Rosas-Casals, M., Corominas-Murtra, B., & Valverde, S. (2008). Robustness of the European power grids under intentional attack. *Physical Review E*. https://doi.org/10.1103/physreve.77.026102.

Trakas, D. N., & Hatziargyriou, N. D. (2018). Optimal distribution system operation for enhancing resilience against wildfires. *IEEE Transactions on Power Systems*, *33*(2), 2260–2271. https://doi.org/10.1109/TPWRS.2017.2733224.

Wang, Y., Chen, C., Wang, J., & Baldick, R. (2016). Research on resilience of power systems under natural disasters—A review. *IEEE Transactions on Power Systems*, *31*(2), 1604–1613. https://doi.org/10.1109/TPWRS.2015.2429656.

Wang, C., Hou, Y., Qiu, F., Lei, S., & Liu, K. (2017). Resilience enhancement with sequentially proactive operation strategies. *IEEE Transactions on Power Systems*, *32*(4), 2847–2857. https://doi.org/10.1109/TPWRS.2016.2622858.

Wang, Z., & Wang, J. (2015). Self-healing resilient distribution systems based on sectionalization into microgrids. *IEEE Transactions on Power Systems*, *30*(6), 3139–3149. https://doi.org/10.1109/TPWRS.2015.2389753.

Wood, E. (2020a). *California opens 'diesel alternative' discussion in microgrid proceeding.*

Wood, E. (2020b). *Batteries, solar, wind, & Australia and Japan: All winners in new Bloomberg NEF forecast.* Microgrid Knowledge.

Xu, Y., & Srivastava, A. K. (2016). Microgrids for service restoration to critical load in a resilient distribution system. *IEEE Transactions on Smart Grid*, *9*(1), 426–437.

York, D., & Jarrah, A. (2020). *Community resilience planning and clean energy initiatives: A review of city-led efforts for energy efficiency and renewable energy.*

Yuan, C. (2016). *Resilient distribution systems with community microgrids.*
Yuan, W., Wang, J., Qiu, F., Chen, C., Kang, C., & Zeng, B. (2016). Robust optimization-based resilient distribution network planning against natural disasters. *IEEE Transactions on Smart Grid*, 7(6), 2817–2826. https://doi.org/10.1109/TSG.2015.2513048.
Zadsar, M., Haghifam, M. R., & Larimi, S. M. M. (2017). Approach for self-healing resilient operation of active distribution network with microgrid. *IET Generation, Transmission and Distribution*, *11*(18), 4633–4643. https://doi.org/10.1049/iet-gtd.2016.1783.
Zerriffi, H., Dowlatabadi, H., & Farrell, A. (2007). Incorporating stress in electric power systems reliability models. *Energy Policy*, *35*(1), 61–75. https://doi.org/10.1016/j.enpol.2005.10.007.
Zia, M. F., Elbouchikhi, E., & Benbouzid, M. (2018). Microgrids energy management systems: A critical review on methods, solutions, and prospects. *Applied Energy*, *222*, 1033–1055. https://doi.org/10.1016/j.apenergy.2018.04.103.

CHAPTER EIGHT

Internet of things and fog computing application to improve the smart-grid resiliency

Janmenjoy Nayak[a], Manohar Mishra[b], Danilo Pelusi[c], and Bignaraj Naik[d]

[a]Department of Computer Science, Maharaja Sriram Chandra Bhanja Deo (MSCB) University, Baripada, Odisha, India
[b]Department of Electrical and Electronics Engineering, FET, Siksha "O" Anusandhan University, Bhubaneswar, India
[c]Faculty of Communication Sciences, University of Teramo, Teramo, Italy
[d]Department of Computer Application, Veer Surendra Sai University of Technology, Burla, Odisha, India

1. Introduction

From the last few decades, the need for ample, clean, and continuous electric energy is expeditiously increasing because of the population growth and climatic changes all over the world (Gungor et al., 2010). To address the issues such as diversification in energy generation, demand response, saving of energy, minimization of carbon footprint, and so on, smart grid is emerging as an important research area in the existing electricity grid. Therefore, the smart grid, which is the integration of conventional electricity grid of the 20th century with the latest information and telecommunication technologies of the 21st century, facilitates the efficient utilization of resources to regulate distributed sources of energy, to exchange the power generated, and to optimize consumption of energy (Anvari-Moghaddam et al., 2015; Han et al., 2014; Wu et al., 2014). As smart grid generates enormous data, intelligent infrastructure is required for fruitful processing of enormous data generated by the smart grid. In this regard, artificial intelligence, internet of things (IoT), cyber physical system (CPS), cloud computing, and fog computing techniques play an important role to develop such infrastructure.

With rising concerns associated with attacks on the cyber and physical assets connected to the smart grid as well as the necessity to mitigate impact of natural disasters, resilience has become a vital and appropriate characteristic. Resiliency is the ability of a system to anticipate and withstand external failure, bounce back to its predefined state as quickly as possible, and adapt to

Electric Power Systems Resiliency
https://doi.org/10.1016/B978-0-323-85536-5.00005-9

be better prepared to future catastrophic events. To improve the resiliency of the smart grid, the system requires continuous monitoring systems well equipped with the above-mentioned intelligent infrastructure.

In this regard, the main intention of this chapter is concentrated on the role of IoT and Fog computing techniques toward the improvement of resiliency in smart-grid infrastructure.

2. Internet of things in smart-grid resiliency

Industrial IoT is one of the important technologies used to support resiliency of smart-grid through an immediate response as soon as a fault occurred. This helps to recover smart-grid networks using real-time monitoring techniques. Because of the extensive usage of IoT in all aspects of the smart grid such as transmission of power, distribution of power, generation of power, and dispatch and utilization of power, extensive research has been carried out on the usage of IoT technology in the deployment of smart grid. Fig. 8.1 shows the growth of publications in smart grid using IoT technology from 2011 to 2021 up to June.

A systematic review on the usage of IoT in smart grid has been highlighted by Reka and Dragicevic (2018). Initially, the review has focused on the characteristics, challenges, and real-time applications of IoT and smart grid. Then, the authors have presented various innovative approaches used

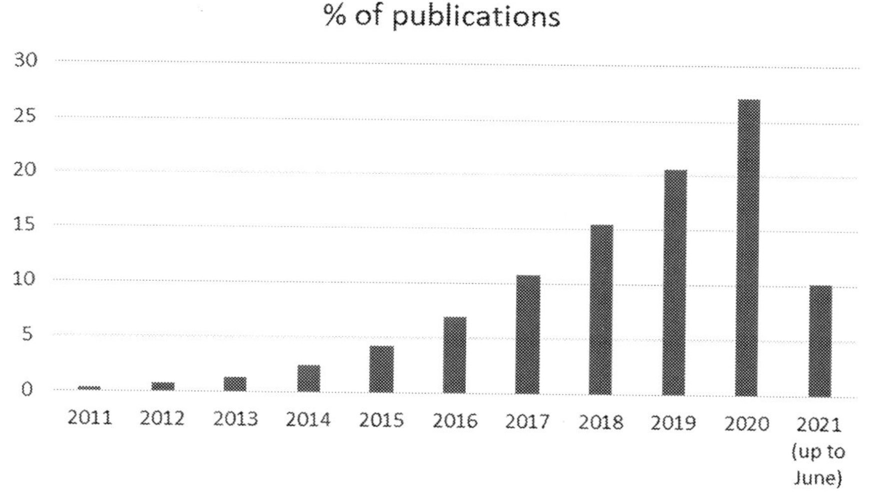

Fig. 8.1 Growth of publications in IoT-based smart grid from 2011 to 2021 up to June. *(No permission required.)*

in the integration of IoT with smart grid along with the benefits and challenges in their corresponding application fields. Moreover, future research directions on IoT-based smart grid have also been presented. To provide a precise prediction of future load value, a novel deep learning system that makes use of IoT technology has been developed by Li et al. (2017) that automatically obtains the characteristics from the captured data. The proposed system makes use of two-step forecasting strategy which automatically enhances the accuracy of the forecasting system. In addition, the proposed system also quantitatively determines the impact of some major factors that provide proper guidance for choosing attribute combination and implementing on board sensors for smart grids with extremely large areas, distinct atmospheric condition, and social rules. Moreover, the experimental results demonstrate that the proposed approach obtains accurate predictions of the future load values. To monitor and control the distribution of electricity in medium and low voltage, an IoT-based smart grid has been proposed by Alharthi et al. (2018). The proposed framework makes use of advancement in the emerging information and communication technology (ICT) technologies, and then incorporates these technologies into the smart grid. Further, the proposed framework makes use of IPv6 routing protocol. Using this IPv6 routing protocol, IoT gateway can be used to associate IP-facilitated smart equipment to the internet. Then, the performance of the proposed IPv6 routing protocol is evaluated by means of average consumption of power, packet delay, and packet delivery ratio (PDR) using COOJA network simulator for Contiki OS. The empirical outcomes reveal that the proposed IPv6 routing protocol performed better in comparison to the conventional RPL-OF0 and RPL-MRHOF protocols. An IoT-based conceptual framework for the smart grid has been developed by Al-Ali and Aburukba (2015). In the proposed framework, each and every device in the grid is considered as object. Further, each object in the grid is allotted a unique IP address using the 6LowPAN communication protocol. Instead of many protocols such as WiFi, Zigbee, WiMax, Bluetooth, PLC, lease lines, and LTE for allocating IP address to the object, the proposed framework uses only 6LowPAN communication protocol. A brief review of the IoT-based smart grid along with security issues and security challenges has been presented by Bekara (2014). A brief survey on the general framework of smart metering and distinct standards of communication and technical research significant to the smart grid has been described by Jain et al. (2014). Furthermore, authors described that consumption of energy can be minimized and efficiency of power can be enhanced by the automation of home, industries, and microgrids using energy management systems. An overview of the demand response in smart grid and the technical research on the demand response in world and in Turkey has been summarized by Elma and Selamoğullari (2017). Furthermore, the authors have conducted a simulation study to reveal the impact of demand response utilization in residential appliances of houses that are expected to be used at high priced hours. Using the smart devices and IoT applications, the demand response will provide advanced perspectives on generation and consumption of electrical energy and on the functioning of power systems. To provide power quality and reliable monitoring, a systematic review on the feasibility of implementing smart meter has been discussed by Anjana and Shaji (2018). Initially, the authors have presented a detailed overview of the smart meters, wireless communication technologies, and routing algorithms which are used as emerging technologies in AMI. Then, the authors have categorized the existing research works based on the target power quality and reliability monitoring. Finally, the authors have outlined future research directions based on the limitations identified in the existing literature. Table 8.1 shows the summary of the literature review of the IoT-based smart grid.

3. Fog computing in smart-grid resiliency

The rapid advancement in the power systems demand the deployment of smart grid which assists in the real-time monitoring and regulating using electricity flows and bidirectional communication. As cloud computing provides parallel processing and sharing of flexible resources and services in the network, it can be used to comply the computational requirements of smart grid applications. Anyhow, the continuous inclusion of distinct sensing and actuating devices, evolution of IoT, and collection of huge amounts of data across the smart grids have led to the intricacy of smart grids. As a result, the architecture of cloud computing is not suitable for providing effective services for smart grid applications. Therefore, fog and edge computing approaches have been used to overcome the problems of cloud computing architecture such as consumption of energy, bandwidth, cost of network, and latency to satisfy the computational requirements of smart-grid applications. As fog computing provides location awareness, analytics of latency, and lower latency to meet the critical requirements of smart-grid applications, several researches have been performed on the deployment of fog computing in smart-grid application. Fig. 8.2 shows the growth of publications in the smart grid using fog computing from 2011 to June 2021.

Table 8.1 Summary of IoT-based smart grid literature review.

Author and year	Focused area	Limitation
Mugunthan and Vijayakumar (2019)	Discussed about the components and solutions of IoT-based smart grids. Also, discussed about the management of home and smart city using smart energy savings. Discussed about the challenges in deployment of IoT-based smart grid along with some solutions to avoid the privacy and security issues	Not focused on discussing the challenges in each layer of framework such as acquisition of energy, communication and congestion in network, handling of enormous data, management of trust, standardization, and so on
Li et al. (2017)	Forecasting of future load values using deep learning framework based on IoT technology	More communication cost due to the transfer of vast amount of data on the communication network
Alharthi et al. (2018)	Monitoring and controlling the electric distributing in medium- and low-voltage networks using IoT-based smart grid and IPv6 routing protocol	Not utilized the proposed IoT-based smart grid and Ipv6 routing protocol in the distribution automation system
Al-Ali and Aburukba (2015)	IoT-based framework for smart grid that makes use of 6LowPAN communication protocol	Conceptual model
Bekara (2014)	IoT-based smart grid along with security issues and challenges	Not focused on the security of advanced metering infrastructure (AMI)

To analyze the trend of research on fog computing in smart grid applications, a systematic analysis from a total of 70 papers has been performed by Gilbert (2019). The study discloses that there is a substantial enhancement in the number of smart-grid applications that make use of the fog and edge computing approaches. Moreover, the study also reveals that a significant number of smart-grid applications are linked with optimization of energy and intelligent organization of the resources of smart grid. Furthermore, it also addresses challenges and issues such as security, programming models, and interoperability, which hamper the smooth utilization of fog and edge computing in applications of smart grid. The motivation for using fog

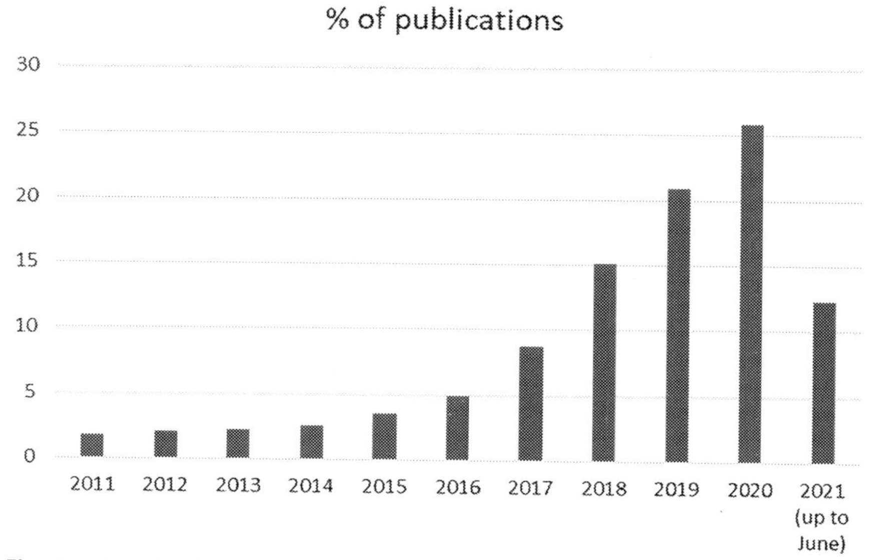

Fig. 8.2 Growth of publications in Fog computing-based smart grid from 2011 to June 2021. *(No permission required.)*

computing as technology in smart-grid analytics has been highlighted by M. Hussain and Beg (2019) by exploring the existing cloud-based smart-grid framework. Then, the authors have developed three-layer fog computing-based smart grid architecture for accommodating number of IoT devices in future smart grids. To determine the association between data consumer, distribution of workload, QoS (quality-of-service) constraints and placement of virtual machines, a cost optimization model for the fog computing has been proposed. As the devised model is an MINLP (mixed-integer nonlinear programming) problem, MDE (modified differential evolution) algorithm is used to solve the problem. Then, the developed prototype has been evaluated using real-time parameters and the results show that 44% of the accumulated power consumption of the cloud computing architecture has been reduced by architecture of fog computing. To overcome the limitations of cloud computing in smart-grid applications, an edge-centered fog computing prototype has been developed by M. Hussain et al. (2019). Initially, the authors have uncovered the deficiencies of the cloud-based smart-grid system. Therefore a prototype has been developed to assure the proper offloading and dispersion of computations across the cloud and edges. In addition, a fog computing SOA (service-oriented architecture) has been developed for IoT-aware smart transportation system operating on the backbone of smart grid. Finally, the challenges that arise in the implementation of the proposed edge-centered fog computing prototype have been highlighted. An extensive analysis of the utilization of fog computing technology in smart grid has been presented by Ruan et al. (2020). Initially, the study focuses on the architecture of smart grid along with its issues and features of fog computing. Then, the study provides a summary of possible solutions provided by fog computing in smart-grid applications and key problems that arise in implementing fog computing in smart-grid applications. Finally, the study focuses on the challenges and future research directions of integrating fog computing technology with smart grid applications. An extensive review on the distinct fog computing approaches that make use of big data along with the existing research on the advancement of smart-grid applications has been presented by Palanichamy and Wong (2019). The summary of literature review of fog computing in smart-grid applications has been illustrated in Table 8.2.

4. Integration of IoT and fog computing in smart grid

From the last decade, smart grid is gaining attention as it enhances the reliability, effectiveness, economics, and viability of electricity services (Bharathi et al., 2017). Presently, the unification of IoT in smart grid plays an essential role in the efficient management of energy infrastructure operations (Goulart & Sahu, 2016). The major challenges encountered in the deployment of IoT in smart grid are the efficient unification and exploitation of ICT infrastructure and smart devices and the processing of huge amount of data produced from the sources (Deng et al., 2017). For efficient handling of large volumes of data, cloud computing technology has emerged, as this technology is capable of providing on-demand availability of computing resources (Rusitschka et al., 2010). However, cloud computing technology was not able to satisfy the critical needs of smart-grid architecture because of the differing modalities of services. Therefore fog computing has emerged to overcome the issues of cloud computing-based smart grid such as connectivity, latency, and bandwidth issues. As fog computing technology can quickly process and interpret the data of IoT applications, several works have been performed on the integration of fog computing and IoT in smart grid. A mathematical model has been suggested by M. Hussain et al. (2020) which describes the planning and placement of fog computing technology in smart grid applications (PPFG). To reduce the average response delay and energy utilization of elements of network, the PPFG model finds

Table 8.2 Summary of literature review of fog computing in smart grid.

Author and year	Focused area	Limitation
Gilbert (2019)	Presented the promises and challenges of fog and edge computing in smart-grid applications. Also, presented the challenges and issues in the implementation of fog and edge computing in smart-grid applications	Not focused on practicality of edge computing in smart grid
A. Kumari et al. (2019)	Uses fog computing-based smart grid framework to make an efficient decision about energy needs of smart devices at the fog layer	Not focused on trading of dynamic energy, privacy, and security perspectives of smart grid
Wang et al. (2018)	Uses a dispersed computing framework that makes use of fog computing technology for IoT applications in the smart grid	Not focused on optimal allocation of resources and not evaluated using communication protocol
Zahoor et al. (2018)	Efficient management of resources in smart grids using cloud-and-fog-based architecture	Not focused on multiple load balancing applications
Barik (2017)	Application of fog computing technology in micro grid	Not focused on distinct types of smart grids that differ in complexity and size

the capacity, optimal location, and number of nodes in fog computing using integer linear programming problem. As the optimization problem is NP-Hard problem, it uses nondominated sorting genetic algorithm to solve the problem. The operation of the suggested model has been demonstrated by executing the PPFG prototype on the smart-grid network. The functioning of design constraints in PPFG model can be better understood by performing a total analysis of the Pareto Fronts which will assist in smart planning and placement of fog computing technology in smart-grid applications. The extent to which cloud computing services can satisfy the significant needs of the smart grid ecosystem have been analyzed by M. Hussain et al. (2018). Simultaneously, the authors have also analyzed the critical needs that can be satisfied in the smart grid through the usage of fog computing utilities. The study also highlights the opportunities and perspectives of using fog computing technology in smart-grid applications. Finally, the study also highlights the challenges and future research directions for the successful transition of smart grid using fog computing. To impart the clearness between the consumer and distributor, an integrated technique based on fog computing and IoT has been presented by Pant et al. (2018), which serves as a backbone for the structure of smart-grid application. In addition, the concept of micro grid, effective load balancing, and power utilization approaches have been introduced along with fog computing to prolong the features to the consumers. To secure the private data of smart meter at the fog server, RSA algorithms have been used. Moreover, the integration of smart street lights with fog computing technology has been done for the effective balancing of load. To effectively plan, supervise, and optimize the activity of household energy, an integrated framework has been suggested by Singh and Yassine (2018). The proposed framework facilitates inventive operation for processing of abundant exquisite data of energy consumption in real life. To focus on the challenges of intricacies and resource needs for real-life processing of data, classification interpretations, and storage, IoT big data analytics structure based on fog computing has been proposed. To completely use the benefits of IoT in smart home, Stojkoska and Trivodaliev (2018) have suggested a three-tier IoT-based hierarchical framework. To combine all renewable dispersed sources of energy from the microgrid and to obtain better optimization, the proposed framework targets to expand the smart home to microgrid level using fog computing. Furthermore, proposed framework has been evaluated using real-time dataset of smart meter and the simulation results show that the suggested approach can minimize the number of transmissions and network traffic for smart home.

5. General discussion and futures directions

To provide resilience, the smart-grid software facilitates customer-side supervision and management of energy utilization. It promotes to a domestic customer a lively contributor in the smart grid's lifespan and administer too. The operation of smart grid depends on wide-ranging technologies and infrastructure way out. IoT-based smart grid includes numerous key components such as smart-grid sensors and meters, automated distribution systems, charging stations, and smart storages. The role of IoT with the integration of cloud/fog computing, big data, and blockchain technologies in smart grid is important. IoT is a well-built enabler of smart grid as its technical and infrastructural mechanisms are for the most part IoT aided. IoT technology in smart grid is extensively expended to automate the operational activities and upsurge efficacy in the supply chain network. By this process, the producers and distributors (i) adopt the advanced metering services to monitor the energy uses in real world and react to mutable demand, (ii) analyze the environmental information to amplify the usage of eco-friendly sources of energy and optimize the power productions from non-conventional energy sources, (iii) monitors the power-grid loads and implements data-driven scheme to minimalize the probabilities of outages/overloads, and (iv) implements the advanced intrusion detection scheme against possible cyber threats.

Big data application in smart grid is also an important development for the power engineer to visualize and analyze the IoT data. Moreover, the application of machine learning (ML) is now common in IoT-based smart grid. We know for a fact that ML is beneficial at working with huge datum datasets. It aids improved understanding and use of big data, recognizing movements, and making predictions. Thus, the application of cutting-edge ML algorithms to analyze IoT data generated in the smart grid supply network is an alternative way to render it beyond capable. Again, the technology named as "Blockchain" offers an exclusive channel for an added decentralized and resilient amalgamation of Energy Internet of Things (E-IoT) and cloud/fog-based gadgets as stated in current researches. Advances in grid-resiliency are vital in the controls and design measures of energy supply networks (ESNs) used to streamline the energy trading. Blockchain-based smart-ESNs can aid these optimization and security gaps and enhance the state-of-the-art in grid resilience through offering an atomically certifiable cryptographic signed distributed register to improve the loyalty, reliability, and resilience of ESNs at the upper hand. The blockchain technology may be expanded to confirm time, customer, and transaction data and protect these records with an absolute crypto initiated register.

5.1 Issues and challenges of IoT-based smart grids

As a CPS, the IoT aid smart grid may encounter the following security issues as mentioned in Table 8.3 (Bekara, 2014). Again to deal with the security

Table 8.3 Security issues of IoT aid smart grid.

Impersonation/identity spoofing	Identity spoofing is an attack having the objective of unauthorized communication through spoofing the identity of smart-grid devices such as smart meter, with an intention to save money for their energy consumption
Eavesdropping	In view of the fact that the smart grid's devices on IoT-based framework can communicate between themself, often working on public communication infrastructure, an attacker can easily have access to their exchanged data
Data tampering	An attacker can amend exchanged data/information
Authorization and control access issues	It is well-known that numerous smart-grid devices are supervised and put together remotely. In this situation, an attacker may possibly try to get illegal access rights to control them, in consequence breaking physical properties or leading to energy outage
Privacy issue	The smart meter or the smart-device used in domestic purpose may store the additional information such as individual habits (such as wake up, sleeping and dinner times, duration of house locked, about vacation, etc.) with the addition to energy consumption data. Thus, by tempering this device, an individual's privacy can be hampered
Compromising and malicious code	The possibility of compromising physically or remotely is high in case of IoT-enabled smart grid as the integrated devices run by some specific computational programme or software. These software may be controlled or manipulated by the attacker through software infection
Cyber-attack	Smart grid is a largest CPS, concerning physical structures representing the physical assets of the smart grid (such as transformers, circuit breakers, smart meters, cables, etc.) and information and communication technology (ICT) systems, where ICT elements control/handle physical objects. Thus, cyberattack could damage the physical properties, which is difficult in the classical power grid

algorithm, protocol and policy for the IoT-aided smart grid, the following mentioned (refer Table 8.4) challenges are important to be taken into considerations (Bekara, 2014). With the above-mentioned security issues and challenges, the key security services needed to improve the resiliency of smart grid are shown in Fig. 8.3.

Table 8.4 Key security challenges of IoT-based smart grid.

Scalability	**As the smart grid spreads over a huge area considering few cities having numerous smart/intelligent devices, it is difficult to ensure scalable security solutions, for example, key management and authentication**
Mobility	Smart grid could involve several mobile devices like electric vehicles or electronic cars and field agents, there will be an endless demand of secure authentication and communication with a mutable circumscribing (such as smart meter and electric charging station)
Deployment	A huge number of smart devices are deployed within the premises of smart grid, which is spread over a large area of a country. Few of them are placed on a remote location with no physical outer limits protections, being easily accessible. It is highly needed to provide accurate security solutions against possible tempering/attack
Legacy systems	The devices or systems that have fixed hardware or software solution and installed on remote islands with no communications or through secret communication infrastructure may face serious challenges when trying to incorporate those legacy systems to IoT-aided smart grid
Constrained resources	In the smart-grid environment, few devices/objects are resource constrained that need special attention during developing their security solutions
Heterogeneity	When dealing with smart-grid devices having diverse characteristics (such as memory, computation, bandwidth, energy autonomy, and time-sensitivity, etc.) and their employed protocols, it is a challenging task to provide secure end-to-end communication. In few cases, the demand of reworking of current solutions is high, even with gateways
Interoperability	In addition to legacy systems and heterogeneity challenges, the interoperability could also be seen as an important challenge of IoT-aided smart grid. Interoperability generally appears when two or more devices working with same protocols and communication stacks but diverse feature capabilities
Bootstrapping	As we know the smart grid includes millions or billions of smart devices, the key challenge is to activate the bootstrapping among them
Trust management	A minimum trust level need to be established whenever we need an efficient communications between the devices used in a smart grid. Therefore, trust management is an important challenge for the utility system

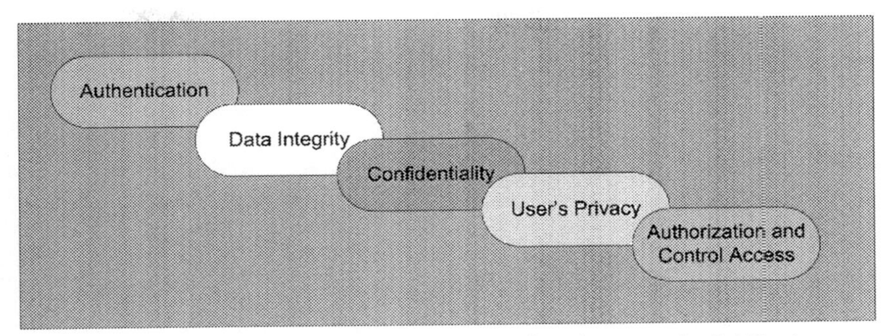

Fig. 8.3 The key security service for the smart-grid resiliency and reliability improvement. *(No permission required.)*

5.2 Issues and challenges of fog computing integration to smart grids

Cloud computing and Fog computing, both are much competitive in nature and offer useful services for data analytics to end consumer. The following details (Table 8.5) provide few key challenges and issues of Fog computing applications in smart-grid environments (Chen & Hao, 2018; A. Kumari et al., 2018, 2019).

6. Conclusion and futures directions

The operational platforms for IoT-based smart grid includes cloud computing, fog computing, and big data. This chapter discusses about IoT and Fog computing in smart-grid applications. To show the current research status in this filed, bibliometric studies are conducted and reported. The outcome of the study demonstrates an exponential growth in the field of security systems in the smart grid over the last decade. IoT-aided smart-grid framework is expected to come into sight rapidly through swollen uses of IoT-based intelligent gadgets like smart meter, sensor, and actuator with an intention to provide complete automation to smart-grid network with better facility of control, protection, and connectivity in the power grid. However, the probability of cyber-security threat in IoT-aided smart-grid framework is high. In this regard, the present research area should be focused on the issues of cyberattack and the development of cyber security mechanisms. The upcoming researches are likely aiming at improving the computational speed of security algorithms though preserving high detection accuracy and a low rate of false alarms. It is also studied that blockchain technology integrated with IoT may help to resolve numerous optimization and

Table 8.5 Key security challenges of fog computing-based smart grid.

Security issues and challenges	Descriptions
Scalability in smart-grid services in real time	It facilitates consumers for real-time supervision of current data uses. Congestion or server breakdown might occur using cloud computing to handle the data analytics facilities and consequently triggers latency
Reliability in smart-grid services	As a result of connection failure and interruption in data processing, cloud computing environments are not consistent in nature; henceforth, maintaining the reliability, scalability, and error-free services are becoming difficult
Personal privacy	Cloud/Fog computing-based platform works on public networks; therefore the security or privacy of data saved on clouds might be compromised. Hence, robust privacy protection schemes against these attackers are necessary to facilitate cloud computing reliability and information security
Connection failure	Connectivity failure is one of the most common issues faced by the consumers. To provide scalability, reliability, and consistency in services, secure multipath Internet transmission channels are required from access centers. Moreover, it is required to increase the utilization of attributes such as multitenancy and virtualization
Heterogeneity in services	The output of smart meters can provide heterogeneous data which are essentially stored in cloud-based platforms for further analysis in future. These data must be accessible by the client whenever required with a nominal time delay. To enable such facilities, smart grid uses cloud; however, connecting diverse categories of intelligent edge gadgets directly to the cloud is a challenging task

reliability challenges that have been accompanied with smart-grid transformation. Fog computing has appeared as a most admired technology of the current industrial revolution. With comparison to cloud computing, it is able to process the vast sized data collected from IoT objects with less processing time. Here, in this chapter, we investigate the current status of Fog computing application in smart-grid ecosystems with the integration of IoT technology. Moreover, it has also studied about the related issues and challenges of Fog computing in smart-grid systems. Fog computing, when accompanied with most advantageous workload allocation schemes, is capable of context-aware support in real-world applications, is responsible for enhanced computational execution performance, and geo-distributed intelligence in smart-grid environments.

Prognostic care is one of the essential issues of smart-grid IoT applications as both upstream and downstream sides power network is built on the usage of costly utensils and infrastructure. Intelligent and smart technologies for supervising and energy management allow stakeholders to control their belongings, forecast failure, and apply well-timed care.

The recent studies reported to Fog computing show its hidden potential as a repository and computational model for emergent Internet-of-Energy ecosystems, for example, smart grid. Considering few additional decisive visions, the real-time experiments are become essential on smart grid architecture. The effectiveness of standard Fog computing model might be upgraded by embedding selective intelligent sensors in the edge nodes. Smart mobility management systems for data-producer and data-consumer will potentially enhance the performance of Fog computing architecture.

Futuristic research scopes need to focus on the state-of-the-art smart-grid implementations in the evolving world. Future researches should concentrate on determining the cost-efficacy of computation methodologies, user-friendliness, data management, scalability, reliability along with security, and privacy in order to improve the decisions of smart-grid engineers. This study can be extended through the inclusion of other academic databases to determine the feasibility of edge computing in smart grid.

References

Al-Ali, A. R., & Aburukba, R. (2015). Role of internet of things in the smart grid technology. *Journal of Computer and Communications*, *3*(5), 229–233. https://doi.org/10.4236/jcc.2015.35029.

Alharthi, S. A., Johnson, P., & Alharthi, M. A. (2018). In *IoT architecture and routing for MV and LV smart grid 2017 Saudi Arabia smart grid conference, SASG 2017* (pp. 1–6). Institute of Electrical and Electronics Engineers Inc. https://doi.org/10.1109/SASG.2017.8356507.

Anjana, K. R., & Shaji, R. S. (2018). A review on the features and technologies for energy efficiency of smart grid. *International Journal of Energy Research*, *42*(3), 936–952. https://doi.org/10.1002/er.3852.

Anvari-Moghaddam, A., Monsef, H., & Rahimi-Kian, A. (2015). Optimal smart home energy management considering energy saving and a comfortable lifestyle. *IEEE Transactions on Smart Grid*, *6*(1), 324–332. https://doi.org/10.1109/TSG.2014.2349352.

Barik, R. K. (2017). FogGrid: Leveraging fog computing for enhanced smart grid network. In *14th IEEE India council international conference (INDICON)*.

Bekara, C. (2014). Security issues and challenges for the IoT-based smart grid. *Procedia Computer Science, 34*, 532–537. Elsevier B.V https://doi.org/10.1016/j.procs.2014.07.064.

Bharathi, C., Rekha, D., & Vijayakumar, V. (2017). Genetic algorithm based demand side management for smart grid. *Wireless Personal Communications, 93*(2), 481–502. https://doi.org/10.1007/s11277-017-3959-z.

Chen, M., & Hao, Y. (2018). Task offloading for mobile edge computing in software defined ultra-dense network. *IEEE Journal on Selected Areas in Communications, 36*(3), 587–597. https://doi.org/10.1109/JSAC.2018.2815360.

Deng, X., He, T., He, L., Gui, J., & Peng, Q. (2017). Performance analysis for IEEE 802.11s wireless mesh network in smart grid. *Wireless Personal Communications, 96*(1), 1537–1555. https://doi.org/10.1007/s11277-017-4255-7.

Elma, O., & Selamoğullari, U. S. (2017). An overview of demand response applications under smart grid concept. In *2017 4th International conference on electrical and electronic engineering (ICEEE)*.

Gilbert, G. M. (2019). A critical review of edge and fog computing for smart grid applications. In *International conference on social implications of computers in developing countries*.

Goulart, A. E., & Sahu, A. (2016). Cellular IoT for mobile autonomous reporting in the smart grid. *International Journal of Interdisciplinary Telecommunications and Networking, 8*(3), 50–65. https://doi.org/10.4018/ijitn.2016070104.

Gungor, V. C., Lu, B., & Hancke, G. P. (2010). Opportunities and challenges of wireless sensor networks in smart grid. *IEEE Transactions on Industrial Electronics, 57*(10), 3557–3564. https://doi.org/10.1109/TIE.2009.2039455.

Han, J., Choi, C. S., Park, W. K., Lee, I., & Kim, S. H. (2014). Smart home energy management system including renewable energy based on ZigBee and PLC. *IEEE Transactions on Consumer Electronics, 60*(2), 198–202. https://doi.org/10.1109/TCE.2014.6851994.

Hussain, M., Alam, M. S., & Beg, M. M. (2018). *Fog computing in IoT aided smart grid transition-requirements, prospects, status quos and challenges*. arXiv preprint arXiv:1802.01818.

Hussain, M., Alam, M. S., & Beg, M. M. (2019). Feasibility of fog computing in smart grid architectures. In *Proceedings of 2nd international conference on communication, computing and networking*.

Hussain, M., & Beg, M. M. (2019). Fog computing for internet of things (IoT)-aided smart grid architectures. *Big Data and Cognitive Computing, 3*(1), 8. https://doi.org/10.3390/bdcc3010008.

Hussain, M. M., Beg, M. M. S., & Alam, M. S. (2020). Fog computing for big data analytics in IoT aided smart grid networks. *Wireless Personal Communications, 114*(4), 3395–3418. https://doi.org/10.1007/s11277-020-07538-1.

Jain, S., Vinoth, K. N., Paventhan, A., Kumar Chinnaiyan, V., Arnachalam, V., & Pradish, M. (2014). Survey on smart grid technologies-smart metering, IoT and EMS. In *2014 IEEE students' conference on electrical, electronics and computer science, SCEECS 2014*IEEE Computer Society. https://doi.org/10.1109/SCEECS.2014.6804465.

Kumari, A., Tanwar, S., Tyagi, S., & Kumar, N. (2018). Fog computing for healthcare 4.0 environment: Opportunities and challenges. *Computers and Electrical Engineering, 72*, 1–13. https://doi.org/10.1016/j.compeleceng.2018.08.015.

Kumari, A., Tanwar, S., Tyagi, S., Kumar, N., Obaidat, M. S., & Rodrigues, J. J. P. C. (2019). Fog computing for smart grid systems in the 5G environment: Challenges and solutions. *IEEE Wireless Communications, 26*(3), 47–53. https://doi.org/10.1109/MWC.2019.1800356.

Li, L., Ota, K., & Dong, M. (2017). When weather matters: IoT-based electrical load forecasting for smart grid. *IEEE Communications Magazine, 55*(10), 46–51. https://doi.org/10.1109/MCOM.2017.1700168.

Mugunthan, S., & Vijayakumar, T. (2019). Review on IoT based smart grid architecture implementations. *Journal of Electrical Engineering and Automation, 1*, 12–20.

Palanichamy, N., & Wong, K. I. (2019). Fog computing for smart grid development and implementation. In *IEEE international conference on intelligent techniques in control, optimization and signal processing, INCOS 2019*Institute of Electrical and Electronics Engineers Inc. https://doi.org/10.1109/INCOS45849.2019.8951412.

Pant, V., Jain, S., & Chauhan, R. (2018). Integration of fog and IoT model for the future smart grid. In *2017 International conference on emerging trends in computing and communication technologies, ICETCCT 2017 (Vols. 2018-)* (pp. 1–6). Institute of Electrical and Electronics Engineers Inc. https://doi.org/10.1109/ICETCCT.2017.8280341.

Reka, S. S., & Dragicevic, T. (2018). Future effectual role of energy delivery: A comprehensive review of internet of things and smart grid. *Renewable and Sustainable Energy Reviews, 91*, 90–108. https://doi.org/10.1016/j.rser.2018.03.089.

Ruan, L., Guo, S., Qiu, X., & Buyya, R. (2020). *Fog computing for smart grids: Challenges and solutions*. arXiv https://arxiv.org.

Rusitschka, S., Eger, K., & Gerdes, C. (2010). Smart grid data cloud: A model for utilizing cloud computing in the smart grid domain. In *First IEEE international conference on smart grid communications*.

Singh, S., & Yassine, A. (2018). *IoT big data analytics with fog computing for household energy management in smart grids* (pp. 13–22). Springer Science and Business Media LLC. https://doi.org/10.1007/978-3-030-05928-6_2.

Stojkoska, B. R., & Trivodaliev, K. (2018). Enabling internet of things for smart homes through fog computing. In *2017 25th telecommunications forum, TELFOR 2017—Proceedings (Vols. 2017-)* (pp. 1–4). Institute of Electrical and Electronics Engineers Inc. https://doi.org/10.1109/TELFOR.2017.8249316.

Wang, P., Liu, S., Ye, F., & Chen, X. (2018). *A fog-based architecture and programming model for IoT applications in the smart grid. ArXiv* https://arxiv.org.

Wu, Y., Lau, V. K. N., Tsang, D. H. K., Qian, L. P., & Meng, L. (2014). Optimal energy scheduling for residential smart grid with centralized renewable energy source. *IEEE Systems Journal, 8*(2), 562–576. https://doi.org/10.1109/JSYST.2013.2261001.

Zahoor, S., Javaid, S., Javaid, N., Ashraf, M., Ishmanov, F., & Afzal, M. K. (2018). Cloud-fog-based smart grid model for efficient resource management. *Sustainability (Switzerland), 10*(6). https://doi.org/10.3390/su10062079.

CHAPTER NINE

Load forecasting using ANN and their uncertainty effect on power system reliability

Subhranshu Sekhar Puhan and Renu Sharma
Department of Electrical Engineering, ITER, SOA Deemed to be University, Bhubaneswar, India

1. Introduction

There is ever-increasing load demand on the present-day power system, which faces highly varying loads on a daily basis. Load forecasting analysis is required to assess the efficiency and performance of power systems. Power utilities are expected to supply dependable capacity to buyers. Their decisions and styles will affect the benefit or loss of crores of rupees for their associations/utilities and also customer satisfaction and future financial interaction in their space. For proficient activity and arranging of the service organization, right load forecasting is fundamental. To give clients continuous and uninterruptible power supply, exact and accurate load forecasting is the fundamental thing for the power system engineers (Moghram & Rahman, 1989). Generally, for short-term load forecasting techniques, various factors are responsible such as weather factors (wind data, dew point, and humidity), time series factors, consumer load usage, and the pattern of load usage by consumers on holiday and working days. Different factors responsible for load forecasting in the power system are discussed in Rothe et al. (2010). There are several types of load forecasting, like very short-term load forecasting, short-term load forecasting, medium-term load forecasting, and long-term load forecasting. The load forecasting techniques vary from some minutes to hours and to years and decades. There are different techniques for load forecasting, out of which multiple regression-based stochastic load forecasting technique is popular (Ruzic et al., 2003). In short-term load forecasting (SLTF), time series factor rule is much crucial. Implementation of load forecasting by time series method such as autoregressive integrated moving average (ARIMA) and autoregressive integrated moving average with exogenous variables has been presented in

Electric Power Systems Resiliency
https://doi.org/10.1016/B978-0-323-85536-5.00002-3

Haida and Muto (1994) and Charytoniuk et al. (1998). A fuzzy auto-regressive moving average with exogenous input variables (FARMAX) at some unspecified time in the future ahead of hourly load forecasting is projected by Yang and Huang (Shayeghi et al., 2015). With concern about neural network and the application of neural network in different fields, with developments of expert systems and different learning methods to NN, short-term load forecasting (SLTF) has been done (Ranaweera et al., 1996). With the advancement of expert system NN and gray system theory (Tayeb et al., 2013), SLTF gives satisfactory results. To get smart forecasting results, day-type data should be taken into consideration. A technique to construct the various ANNs for everyday type and feed every ANN with the corresponding day type training sets was discussed in Khotanzad et al. (1997). The effects of load forecast uncertainty in power system reliability assessment incorporating changes in system composition, topology, and load curtailment policies have been discussed by Abdel-Karim et al. (2014). The next section of this chapter is organized as follows, purpose of load forecasting and classification is described in Section 2. Section 3 deals with ranking of input data by analysing performance index, Section 4 deals with different ANN Based load forecasting techniques. In this manuscript Section 5 describes the LEVENBERG-MARQUARDT solution methodology, Sections 6 and 7 deals with Load forecasting uncertainties and their effect on power system reliability and results respectively. The last section in this manuscript deals with conclusions.

The main contribution of authors toward this chapter is as follows:

- Application of accurate load forecasting using ANN techniques with sensitive weather data.
- Ranking the weather data according to performance index based on Fuzzy logic
- Select the top-ranking data for ANN application and remove the redundant set of data.

2. Purpose of load forecasting and classification

The need of load forecasting in power systems basically deals with the following:

- Power system planning.
- Exact information about the load.
- Power grid expansion.
- Marketing strategy of power grid.
- Load shedding curtailment.

The forecasting techniques can be classified into mainly three types, extrapolation, correlation, and a combination of both extrapolation and correlation methods used. The above-mentioned forecasting techniques can be differentiated into deterministic, probabilistic, and statistical methods. The extrapolation method generally deals with curve fitting techniques, whereas the correlation approach generally relates the relationship between the dependent and independent variables. The correlation approach is most feasible as compared with the extrapolation method because in correlation methods the forecaster may get a rough idea between the load forecasting and the measurement variables. Classification of load forecasting can be classified into four types such as:

- Very short-term load forecasting
- Short-term load forecasting (STLF)
- Medium-term load forecasting (MTLF)
- Long-term load forecasting (LTLF)

The load forecasting techniques can be classified into different categories according to the prediction of load in different time intervals (Table 9.1). If the load is forecasted in a short span of time for let minute basis, then it is a very short-time load forecast, and if it is on an hourly basis, then it is called short-time load forecasting. In long-term forecasting, the load is just forecasted for some year basis or on decade basis also. Within the deregulated economy, the behavior of the consumer, weather prediction data, and the holiday profile of the year present a most valid discussion for short-term load forecasting (SLTF) and very short-term load forecasting. Predicting load forecasting in quick succession is a cumbersome task for power system Engineers.

Nowadays, researchers are more focused on ANN-based methods and support vector-based methods to predict the SLTF. The advancement of statistical learning theory enhances the predictability rate of support vector machines. Generally, the end-user model and econometric model are followed by the power system Engineers to study the medium-term and long-term load forecasting methods. In the end-user model, various factors such as customer use, size, and age group of customers are generally

Table 9.1 Categorization of load forecasting.

Nature of forecast	Forecast interval	Application
STLF	Seconds, minutes, hours	Unit commitment
MTLF	Few days, weeks	Seasonal peak
LTLF	Year basis, sometimes decades	Power system planning

considered. In the econometric method, least-square error method and other statistical method are used to interrelate the dependent and independent variable.

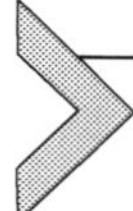

3. Relative ranking of input data by analyzing performance index

To reduce the effect of masking and the mis-ranking, generally Fuzzy logic-based performance index is used. Here, we have used Fuzzy logic-based performance index based on two objective criteria based on wind speed and humidity. Table 9.2 shows the relation between linguistic variable and wind speed variation and Table 9.3 shows the relation between linguistic variable and humidity level.

The performance index in a general term can be written as mentioned in Prasad et al. (2020)

$$PI = \sum_{i=1}^{n} \frac{W_i}{2n} f(Z)^{2n} \tag{9.1}$$

where $f(Z)$ is a linear function, which depends on the square of the difference of actual load and specified load in a particular slot and W_i denotes weight factor combined in each slot.

$$Dw\ d_i = \frac{2\Delta wd_i}{wd_{imax} - wd_{imin}} \tag{9.2}$$

where Dwd_i, wd_{imax}, wd_{imin} are wind deviation in ith slot, maximum and minimum wind deviation, respectively, from base speed.

The performance index related to wind speed will be calculated according to Eq. (9.3).

Table 9.2 Relation between linguistic variable and wind speed variation.

Linguistic variable	S	M	L	VL
Dwd	0–0.2	0.2–0.6	0.6–1.04	1.04+

Table 9.3 Relation between linguistic variable and humidity level.

Linguistic variable	S	M	L	VL
MI	0–0.24	0.24–0.6	0.6–1.4	1.4+

$$P\,I_{wd} = \sum_{i=1}^{n} \frac{W_i}{2n} \times Dw\,d_i \tag{9.3}$$

These normalized deviations are arranged in a Fuzzy logic network like mentioned in the below table using fuzzy set notations like S (small), M (medium), L (large), and VL (very large).

To include the effect of humidity, the following equation is used:

$$(NMA)_i = \frac{(MH)_i - H_b}{(MH)_B - H_b} \tag{9.4}$$

where

$(NMA)_i$ is the normalized humidity level,

MH_i and MH_B are maximum humidity in slot i and maximum humidity in base case.

H_b signifies humidity in base case

$$MI = 1 - NMA \tag{9.5}$$

$$PI_H = K \times MI \tag{9.6}$$

Total performance index can be calculated as,

$$P\,I_{total} = P\,I_{wd} + G \times P\,I_H \tag{9.7}$$

4. ANN-based load forecasting techniques

ANNs are modern artificial intelligence models that solve nonlinear functions, data sorting, pattern recognition, optimization, prediction, modeling, system identification, forecasting, management, and simulation very well. ANN-based techniques give the best result when there are uncertainties in the model and when there are nonlinearity dynamics present in the situation. As the present demand for load forecasting is of an uncertain nature, the application of ANN in load forecasting (SLTF) is much suitable. There are various parameters responsible for the design of ANN, like the selection of hidden layers, the selection of input parameters, and the training method. The number of hidden layers off course enhances the predictability of the data set but choosing the number of hidden layer clearly depends on the data set used. Here, we have used one hidden layer, and one input and output layer both, that is, we have used multilayer perceptron-based ANN. The activation function used can also play a significant role in the ANN design procedure. To illustarate such multilayered feed-forward networks possessing, an auxiliary layer is considered between the input layer and the output layer. The hidden neurons present within the middle layer were considered for computational purposes.

5. LEVENBERG-MARQUARDT solution methodology

There are generally five steps to be followed for LEVENBERG-MARQUARDT training purposes.

1. Data collection.
2. Data handling and preparation.
3. Network structure design.
4. Training method used for training purposes.
5. Validation of the training data.

In the data preparation stage, to free the data from any bias present in the input data, the data need to be processed by scaling the input data set and setting the various parameters like mean standard deviation and a central tendency for the training data set. In the network structure design stage, multilayer feed-forward techniques are used. The number of the input node used is generally decided by the number of input parameters taken into study. Here, we have taken the load historical data, the wind data, and the humidity data near to Dadri, Delhi area (refer Humidity with load data, n.d.; Wind data, n.d.); hence, the total number of nodes used in input is three (3).

The next and final step in the ANN design is to training the network, and to just do away with changing the association of weights. Training of an ANN network is just an iterative solution to change the associated weight in a comprehensive manner. LEVENBERG-MARQUARDT is an optimized solution method to optimize the performance indices associated with weights.

The performance index used for the LEVENBERG-MARQUARDT training method is:

$$F(w) = \sum_{p=1}^{p} \left[\sum_{k=1}^{k} \left(d_{kp} - O_{kp} \right)^2 \right] \quad (9.8)$$

where d_{kp}, O_{kp} stands for the measured and standard weight between k and P node.

Levenberg Marquardt algorithm consolidates the speed of Gauss-Newton's method and also the stability of the error back propagation algorithm during training.

The update in weight in the next layer is as follows:

$$\Delta w = \left(\left(J^T J + \mu w \right)^{-1} \right) + J^T e \tag{9.9}$$

where J is the Jacobian matrix and μ is the learning rate parameter, e is the error.

Once the μ value is small, the weight updation follows the Gaussian method, and if μ is large, it follows Newton's method.

$$J = \begin{bmatrix} \frac{\partial F(x_1, w)}{\partial x_1} & \frac{\partial F(x_1, w)}{\partial w} \\ \frac{\partial F(x_n, w)}{\partial x_n} & \frac{\partial F(x_1, w)}{\partial w} \end{bmatrix} \tag{9.10}$$

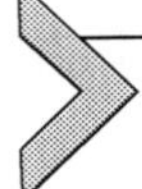

6. Effect of load forecast uncertainties on reliability of power system

Load forecast uncertainty can be described by a probability distribution whose parameters can be estimated from past experience and future considerations. To accommodate probability distribution function, as it is difficult to obtain comprehensive information about the historical load data, normal distribution of continuous random variables is preferred. The approximation value of load distribution can be obtained by tabulating the normal distribution or by just applying interval method.

Some of the researchers are also introducing Monte Carlo simulation packages to test the effect load forecast uncertainties in bulk electric power system.

The tabulating procedure for introducing normal distribution to the probability distribution and the sampling methods are the best described methods in literature (Abdel-Karim et al., 2014). In the tabulating method, the probability distribution function is divided into P equal intervals.

$$F(x_i) = \frac{i - 0.5}{P} \tag{9.11}$$

where x_i is a random number which is normally distributed having mean of 0 and standard deviation of 1.

$$x_i = F^{-1}\left(\frac{i-0.5}{N}\right) \tag{9.12}$$

$$x_i = \begin{cases} -Z & 0.5 > x_i \\ 0 & x_i = 0.5 \\ Z & x_i \geq 0.5 \end{cases} \tag{9.13}$$

$$z = w - \frac{\sum_{i=0}^{2} c_i \times w_i}{1 + \sum_{i=1}^{3} c_i \times w_i} \tag{9.14}$$

$$w = \sqrt{-2\ln q} \tag{9.15}$$

$$q = \begin{cases} 1 - F(x_i)\ 0.5 \leq F(x_i) < 1 \\ F(x_i)\ 0 < F(x_i) < 0.5 \end{cases} \tag{9.16}$$

$$RMSE = \frac{1}{\sqrt{N}} \sqrt{\sum_{i=1}^{N} (load_{actual} - load_{Forecast})^2} \tag{9.17}$$

The weight of the value of c and d coefficients are taken from Abdel-Karim et al. (2014).

7. Result and discussion

Wind speed deviation and humidity deviation performance indexes are ranked for different slots, and the loads are predicted for those slots using LEVENBERG-MARQUARDT-based algorithm. To predict the load for the different slots according to the ranking, the sample weather and load data of May and June months of the past 6 years (2008–13) are used. For that, equivalent days of Friday are identified to be Monday, Tuesday, Thursday, and Friday. So, in total, we have 255 data points. Of these data points, 80% have been used for the training purpose, 10% for validation, and the remaining 10% for testing. Tables 9.4 and 9.5 depict the ranking of various slots with performance indices for wind and humidity data obtained from Fuzzy logic network, respectively, and the details of total performance index (mentioned in Table 9.6) is calculated using Eq. (9.9). The wind speed data are taken from the Indian Metrological department and the load profile data with humidity is taken from the SLDC Delhi region for the time frame. Different slots of humidity and wind data are taken along with load and Fuzzy logic linguistic variable used to eliminate the less significant data.

The probability uncertainties are calculated using Eqs. (9.7)–(9.9) mentioned above for five different slots as depicted in Eqs. (9.11)–(9.16). The

Table 9.4 Ranking of various wind slot with performance index.

Rank	Slot No.	
1	25	499.778
2	36	396.824
3	28	259.333
4	40	147.356
5	37	142.556

Table 9.5 Ranking of various humidity slot with performance index.

Rank	Slot No.	
1	25	329.738
2	37	286.224
3	40	052.163
4	41	046.389
5	36	042.245

Table 9.6 Ranking slot with total performance index.

Rank	Slot No.	
1	25	309.632
2	37	316.124
3	41	252.123
4	40	142.089
5	32	142.147

Table 9.7 Probability of uncertainties in load forecasting.

Slot	$i=0$	$i=1$	$i=2$	$i=3$	$i=4$
Random number	1	2	3	4	5
Probability	$p_i = 0.086$	$p_i=0.09814$	$p_i = 0.092$	$p_i = 0.029$	$p_i = 0.708$

probability of uncertainties for five different slots according to ranking is illustrated in Table 9.7.

Fig. 9.1 shows the error histogram with 20 bins. The top 20 ranks are tested with the help of ANN and the error histogram details are plotted.

The gradient and rate of learning for different epoch and also the validation check of all epoch used in ANN tool box is plotted in Fig. 9.2. Similarly, the training, testing, validation, and overall three of the phases is

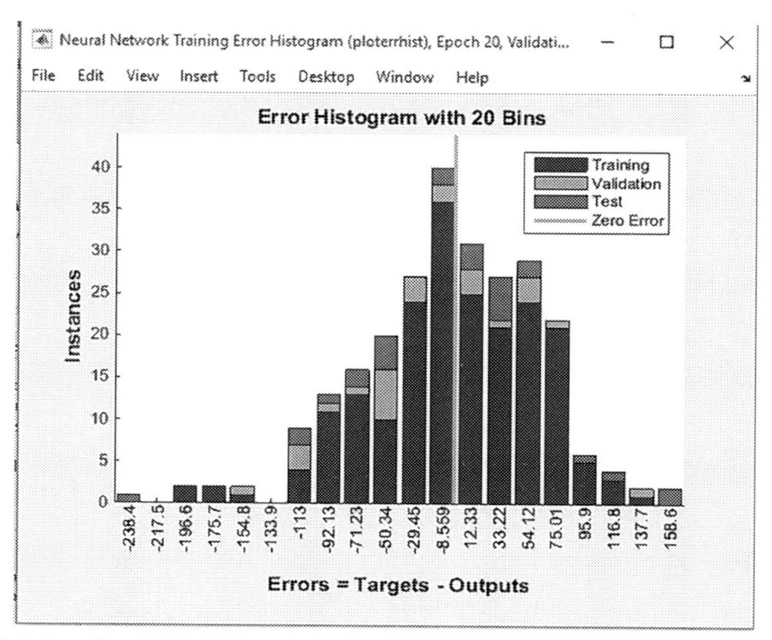

Fig. 9.1 Error histogram with 20 ranks. *(No permission required.)*

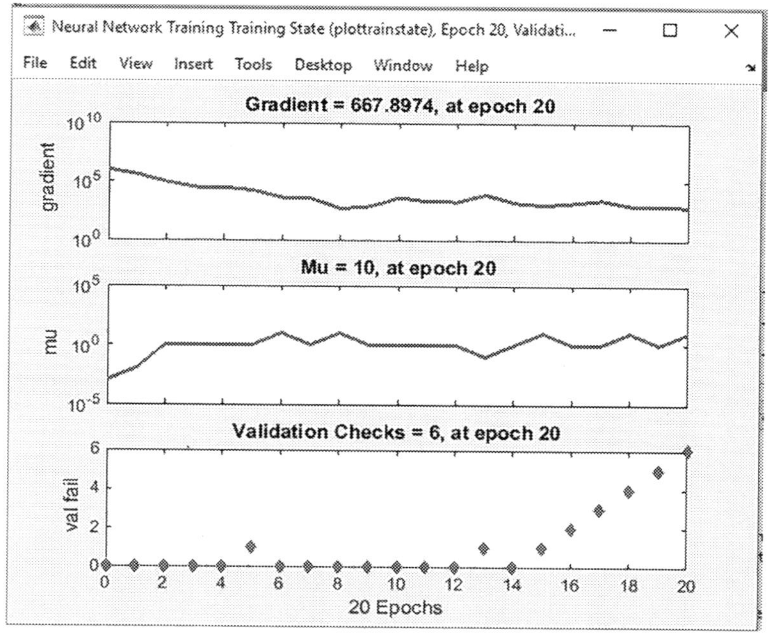

Fig. 9.2 Gradient, rate of learning and validation check plot for different epoch. *(No permission required.)*

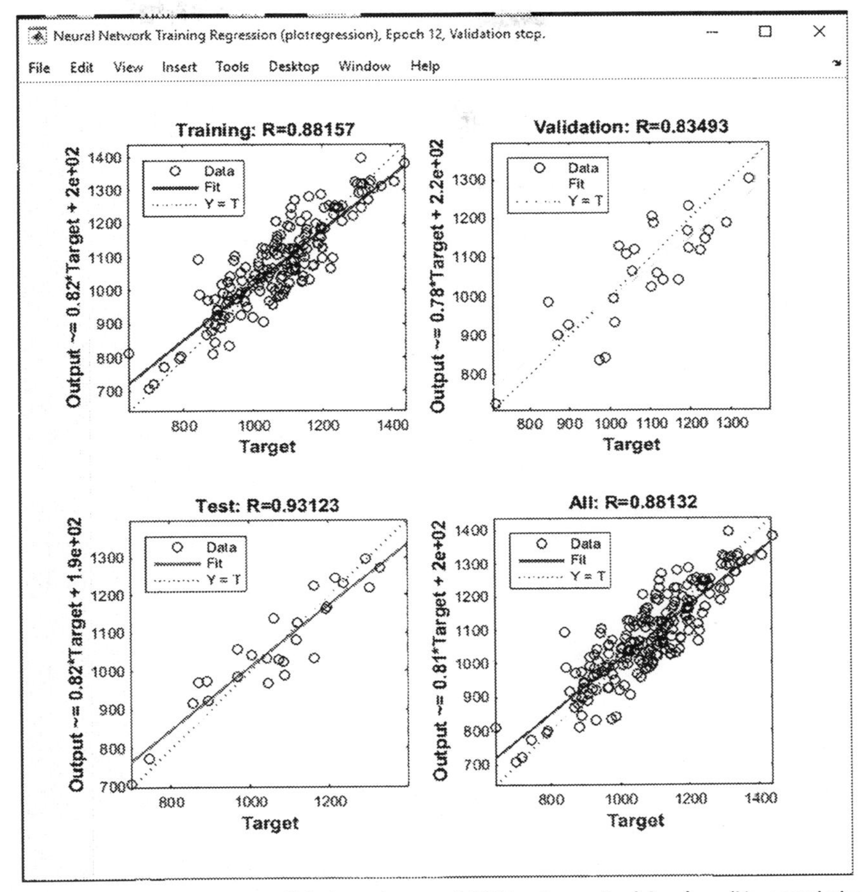

Fig. 9.3 Training, testing, validation phases of ANN using suited R value. *(No permission required.)*

plotted with regression analysis. The best suited regression coefficient is found to be around 0.88132 as depicted in Fig. 9.3.

Finally, the actual load on different slot ranked according to Fuzzy logic-based performance index along with RMSE error is tabulated in Table 9.8. The RMSE value is calculated after getting the forecasted data and substituting the same in Eq. (9.17).

Table 9.8 Load forecasted for different slot with RMSE.

Rank	Slot	Actual load (kW)	Forecasted load (kW)	RMSE (%)
1	25	977	980	0.65
2	37	908	895	2.34
3	41	969	1002	8.57
4	40	965	956	1.19
5	36	896	889	0.12

8. Conclusion

Short-term load forecasting (STLF) has been predicted for a 15-minute slot, which is according to the ranking done by Fuzzy logic based Performance index. This chapter discusses the STFL prediction along with RMSE error by allocating a normal PDF function. The load data are predicted by taking the weather prediction data (wind and humidity data near Dadri region). The reliability analysis of STFL uncertainties on bulk electric power systems are also discussed by using some normal distribution-based PDF function.

References

Abdel-Karim, N., Nethercutt, E. J., Moura, J. N., Burgess, T., & Ly, T. C. (2014). Effect of load forecasting uncertainties on the reliability of North American bulk power system. In (Vols. 2014). *IEEE Power and Energy Society general meeting, October*IEEE Computer Society. https://doi.org/10.1109/PESGM.2014.6939284.

Charytoniuk, W., Chen, M. S., & Van Olinda, P. (1998). Nonparametric regression based short-term load forecasting. *IEEE Transactions on Power Systems, 13*(3), 725–730. https://doi.org/10.1109/59.708572.

Haida, T., & Muto, S. (1994). Regression based peak load forecasting using a transformation technique. *IEEE Transactions on Power Systems, 9*(4), 1788–1794. https://doi.org/10.1109/59.331433.

Humidity with load data. (n.d.). Retrieved from: http://www.delhisldc.org/Redirect.aspx?Loc=0804. (Accessed 13 December 2020).

Khotanzad, A., Afkhami-Rohani, R., Lu, T. L., Abaye, A., Davis, M., & Maratukulam, D. J. (1997). ANNSTLF - A neural-network-based electric load forecasting system. *IEEE Transactions on Neural Networks, 8*(4), 835–846. https://doi.org/10.1109/72.595881.

Moghram, I., & Rahman, S. (1989). Analysis and evaluation of five short-term load forecasting techniques. *IEEE Transactions on Power Systems, 4*(4), 1484–1491. https://doi.org/10.1109/59.41700.

Prasad, M., Puhan, S., & Patra, S. (2020). Contingency selection in the context of voltage security margin determination. *ICICCT 2019 – System Reliability, Quality Control, Safety, Maintenance and Management: Applications to Electrical, Electronics and Computer Science and Engineering*. Springer, 10.1007/978-981-13-8461-5_85.

Ranaweera, D. K., Hubele, N. F., & Karady, G. G. (1996). Fuzzy logic for short term load forecasting. *International Journal of Electrical Power and Energy Systems, 18*(4), 215–222. https://doi.org/10.1016/0142-0615(95)00060-7.

Rothe, J. P., Wadhwani, A. K., & Wadhwani, S. (2010). Hybrid and integrated approach to short term load forecasting. *International Journal of Engineering Science and Technology, 2*, 7127–7132.

Ruzic, S., Vuckovic, A., & Nikolic, N. (2003). Weather sensitive method for short term load forecasting in electric power utility of Serbia. *IEEE Transactions on Power Systems, 18*(4), 1581–1586. https://doi.org/10.1109/TPWRS.2003.811172.

Shayeghi, H., Ghasemi, A., Moradzadeh, M., & Nooshyar, M. (2015). Simultaneous day-ahead forecasting of electricity price and load in smart grids. *Energy Conversion and Management, 95*, 371–384. https://doi.org/10.1016/j.enconman.2015.02.023.

Tayeb, E. M., Ali, A., & Emam, A. A. (2013). Electrical energy management and load forecasting in a smart grid. *International Journal of Engineering Inventions, 2*(6), 98–101.

Wind data. (n.d.). Retrieved from: http://mausam.imd.gov.in. (Accessed 13 December 2020).

CHAPTER TEN

Cost-benefit analysis for smart grid resiliency

Sonali Goel and Renu Sharma
Department of Electrical Engineering, Institute of Technical Education and Research, Siksha 'O' Anusandhan (Deemed to be University), Bhubaneswar, India

1. Introduction

Apart from the growing concern about the critical need for improved resilience in the energy sector, there is no uniform framework for assessing resiliency levels or exploring potential improvement options. Although high-and low-frequency events cannot be prevented, the goal of resiliency damage prevention, recovery, and survival is to reduce the associated costs. Governments and public authorities use the cost-benefit analysis method to measure societal benefit for investment in different projects and policies. Typically, it is used to plan and present costs and benefits, as well as inherent tradeoffs, for estimating the economic efficiency of a project (Proag & Proag, 2014). Before building or undertaking a new project, prudent managers undertake a cost-benefit analysis to assess all of the project's possible expenses and revenues. The study's findings will decide whether the project is financially viable or whether the company should pursue another initiative. Cost-benefit analysis (CBA) is important during project design to detect early implementation issues such as cost and financial feasibility, as well as to see whether an alternative design might have a larger impact per capita spent. CBA models generate projections over some time of analysis of 10–20 years from the start of a project/activity, which is important for determining the investment's long-term viability. This chapter describes the influence of resilience on the economics of solar photovoltaic system systems for commercial buildings.

1.1 What is resiliency and why it is important

Resiliency is the ability of the power system to withstand, respond, and recover from a catastrophic event. These types of events can be caused by environmental threats and human threats, such as cyber security attacks.

Electric Power Systems Resiliency
https://doi.org/10.1016/B978-0-323-85536-5.00008-4

The electric power system resiliency is threatened by extreme weather events and natural catastrophes, such as hurricanes and tornadoes, flooding and sea level rise, earthquakes and consequent tsunamis, severe wildfires, drought, or heatwaves, and severe wind storms. Recent severe weather events have wreaked havoc on electric infrastructure, leaving consumers without electricity and, in some cases, without clean water, food, or fuel for heating or transportation. A hazard is an actual exposure to something of human importance to a threat and is often considered as a combination of probability and loss. It is a potential threat to human and their wellbeing, and the probability of such occurrence is known as risk (Proag & Proag, 2014). Resiliency is an important concept in finding a flexible framework for benefit-cost analysis that can assist in assessing and prioritizing investments.

This chapter is organized into different sections. Section 1 presents the introduction to resiliency, whereas Section 2 presents the elements of a resiliency framework. The cost-benefit analysis tool is presented in Section 3. Section 4 presents the approach for cost-benefit analysis of a project. Section 5 describes the cost-benefit analysis advantages, whereas Section 6 discusses the use of resiliency in a project. The economic analysis of a PV system is discussed in Section 7 followed by a conclusion in Section 8.

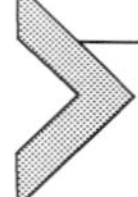

2. Elements of resilience framework

The elements of a resilience framework are shown in Fig. 10.1.

2.1 Cost-benefit analysis framework

Cost-benefit analysis is a methodology used by companies to estimate the likely costs and benefits of potential projects. CBA uses several tools for addressing uncertain outcomes and values, including sensitivity, probability, and break-even analysis (Makowsky & Wagner, 2009). There are four approaches used to determine the CBA. These are Engineering Estimate, Parametric Modeling, Analogy Estimating, and Delphi Method. Parametric Modeling and Engineering Estimates are generally used in an engineering environment. Both of these techniques share the same approach (Misuraca, 2014). The cost-benefit analysis framework is shown in Fig. 10.2.

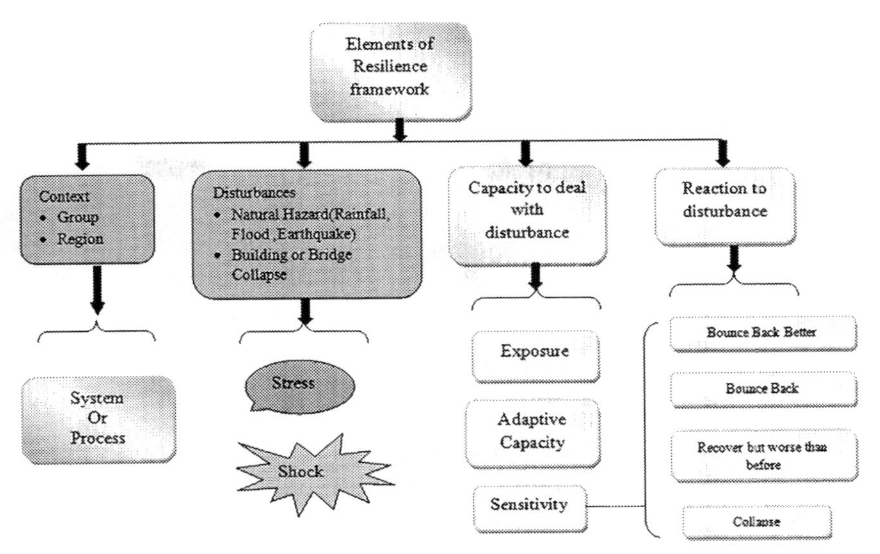

Fig. 10.1 Elements of a resilience framework. *(No permission required.)*

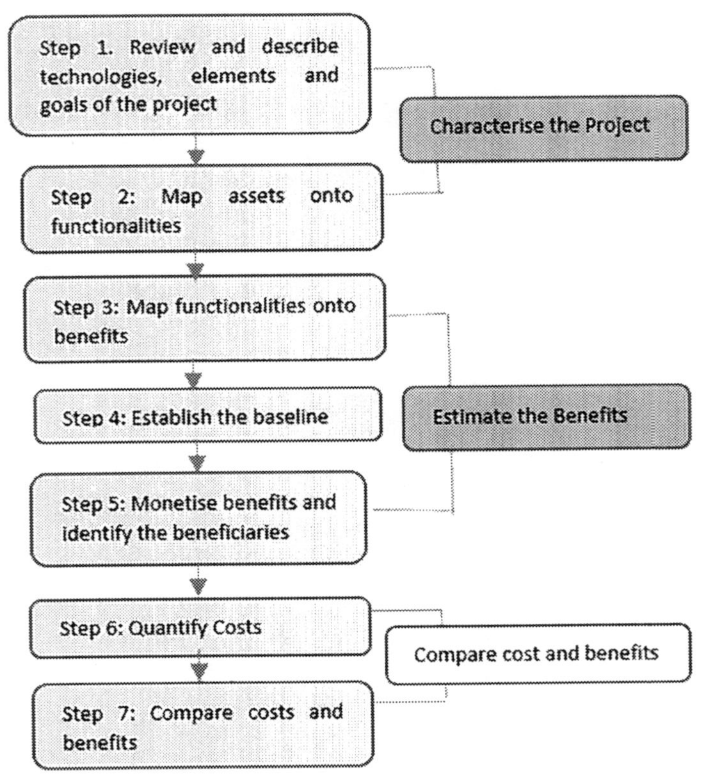

Fig. 10.2 Cost-benefit analysis framework. *(No permission required.)*

3. Cost-benefit analysis tool

Cost-benefit analysis is one of the most important tools for performance appraisal. Cost-benefit analysis is best suited for small to medium projects that do not take long to complete. In these cases, analysis can guide those involved in making the right decisions. However, large-scale long-term projects can be problematic in terms of cost-benefit analysis (CBA). Cost-benefit analysis in project management aims to provide an orderly approach to find out the pros and cons of a project, including investments, activities, market needs, and expenses. CBA provides an alternative to determine the best approach for achieving the target while saving on the investment. CBA has two primary purposes:

To decide whether the project is sound, justifiable, and viable by deciding whether its benefit prevails over cost.

To provide a benchmark for project evaluation by assessing the advantages of which project are greater than its costs.

Cost-benefit analysis as a tool is shown in Fig. 10.3.

There are other factors, such as inflation and interest rates, that affect the accuracy of the analysis. There are other ways to complement the CBA in evaluating major projects, such as net present value (NPV) and internal rate of return (IRR). Compiling a detailed list of all the costs and benefits connected with the project or choice should be the first step in a cost-benefit analysis. However, the use of CBA is an important step in deciding which project to pursue. Once the cost and benefit have been calculated, there are

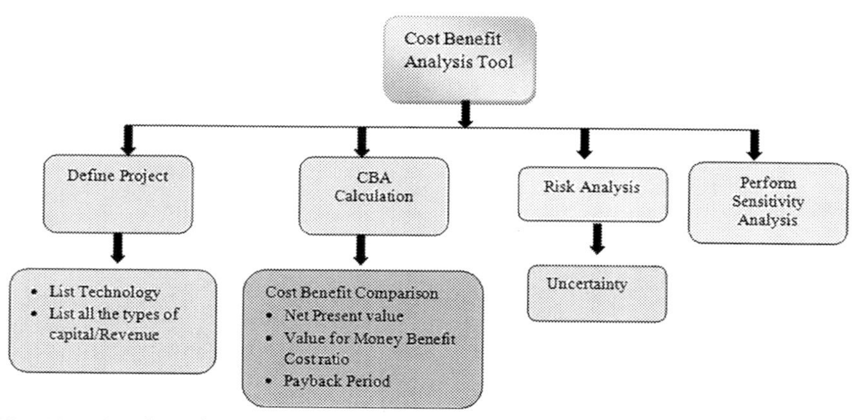

Fig. 10.3 Cost-benefit analysis as a tool. *(No permission required.)*

various ways to compare them to determine the project's cost effectiveness. In a conventional benefit-cost analysis, investments that can be evaluated by comparing the costs and benefits expressed in the present value method make the comparison more accurate. If the costs or benefits are not known with certainty, then the analysis should account for this in terms of expected risk (Landau, 2021).

4. Cost-benefit analysis approach

There are different approaches for assessing a project's cost-benefit analysis. They are:

- Define project

  The initial costs of each of the measures proposed are examined in this chapter. Although designing and building a more robust system will always cost more, this initial investment may result in cost savings in the long run. Reduced operation and maintenance costs, decreased repair costs, and shorter system downtimes, for example, might all contribute to lower lifespan costs. Although the focus is on extreme weather zones, many of the design concepts might help other areas become more resilient.

  Various approaches for conducting cost-benefit analysis are shown in Fig. 10.4.

- What are the project goals and objectives?

  The first step is the most critical, because a clear and concise idea is required before we can determine whether a project is worth the effort.

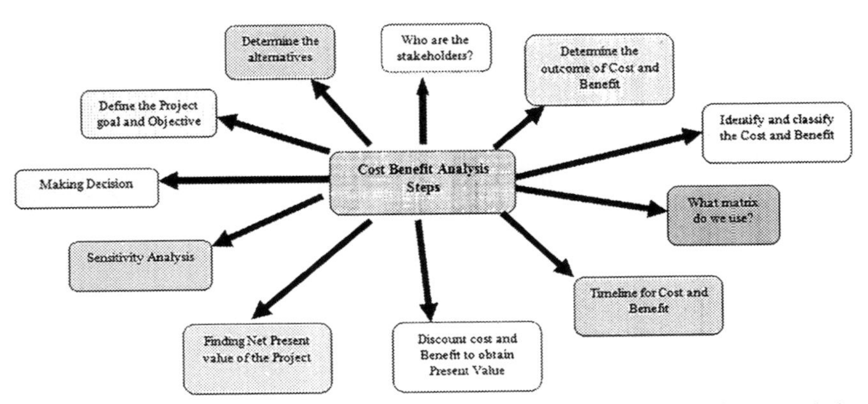

Fig. 10.4 Various approaches in conducting cost-benefit analysis. *(No permission required.)*

We need a straightforward and brief understanding of what we are going to do before we can determine whether a project is worth the effort.

- What are some alternatives?

  We need to equate the existing project with other projects before we decide whether a project is right and see which way to go.

- Who are the stakeholders?

  We have to find out the list of all the stakeholders in the project.

- What measurements do we use?

  For calculating all costs and benefits, we need to decide about the matrices that we will use. Also, how are such metrics going to be reported? We can generate multiple project reports with a single click, including project status reports, variance reports, and many more, with the "ProjectManager.com" website.

- To find out the outcome of the cost and benefit.

  We have to find out what are the total costs and what are the benefits of doing a project over a significant period.

- What is the discount rate?

  The discount rate indicates the changing value of money over time. It is equivalent to the amount of money that we could make from the capital if we invest it in a bank or other financial investment other than investing in power plants. It represents how much any future amount is discounted or reduced to make its corresponding equal amount today. The discount rate is similar to an interest rate but may not be equal to the interest rate.

- What is the NPV of a project?

  The NPV of a system is the present value of all the investment costs that it incurs during its lifetime minus the present value of all the revenue that it earns over its lifetime. Costs include capital costs, replacement costs, O&M, fuel costs, emissions penalties, and the costs of buying power from the grid. Revenues include earning from salvage value and from electricity sales to the grid; it is given by Lin et al. (2011). In other words, NPV is calculated by subtracting the present values of all investment costs from the present values of all benefits (revenue) (Sowe et al., 2014).

NPV of the project is given by (Goel et al., 2021)

$$NPV = -S + \frac{CF_1}{(1+r)^1} + \frac{CF_2}{(1+r)^2} + \cdots + \frac{CF_n}{(1+r)^n} = -S + \sum_{j=1}^{n} \frac{CF_j}{(1+r)^j}$$

where

S = capital investment of PV system.

r = annual discount rate.

CF_j = net cash inflow in the year j.

Simply, it is calculated by (Goel et al., 2021)

$$NPV = -\text{Initial cost} + PV\,[\text{Electricity income} - \text{Operating and main cost} - \text{Inverter and component replacement} + \text{Salvage value}]$$

In other words, NPV is calculated by subtracting the present values of all investment cost from the present values of all benefit (revenue).

$$NPV = \text{Present value of benefit} - \text{Present value of cost}$$

NPV of a project indicates the acceptance criterion, and the benefit earned should always be greater than investment costs. The NPV should always be a positive value.

- Benefit-cost ratio

  Risk to resilience project mainly focuses on the benefit to cost (B/C) ratio or BCR. Benefit-cost ratio is the ratio of the total present value of the benefit to the present value of all cost investment. If the BCR is greater than 1, the project is considered a profitable venture. A benefit-cost ratio can be used to describe the overall relationship between a proposed project's relative costs and benefits.

- Payback period

  Payback period (PBP) is the number of years required for an investment to be recovered from the net cash flow from benefit.

$$\text{Payback period} = \frac{\text{Total cost of the Plant}}{\text{Net benefit per year}} \tag{10.1}$$

- Risk analysis

  Risk transfer is a risk management and control strategy that involves the contractual shifting of a pure risk from one party to another. Risk transfer measures typically require a consistent annual payment, for example, insurance premium guaranteeing financial security in case of an event. These costs can usually be determined directly because market prices exist for cost items such as labor, materials, and other inputs. Some uncertainty in this estimate usually remains as the price for inputs and labor may fluctuate. Most often, project appraisal documents make allowances for such potential fluctuations by varying cost estimates by a certain percentage when appraising the costs.

Key information needed for the detailed cost-benefit analysis on risk management measures includes (i) The exact type of the option under consideration, (ii) its expected lifetime, (iii) costs of investment and its operating costs, (iv) planned sources of financing, (v) additional future benefits and its impact.

- Perform sensitivity analysis

  Sensitivity analysis is a common method used to measure the risk of the project by identifying the critical input parameter. The main step in conducting a sensitivity analysis is to define the objective of the problem followed by the identification of which parameters will be taken into account.

- What we should do?

  After collecting all the data, the final step is to make a recommendation based on analysis.

5. Cost-benefit analysis: Advantages

Cost-benefit analysis is a popular tool with the following advantages (Cost-benefit analysis: Advantages, limitations, examples, and relevance, 2019).

1. Complex decisions of a project can be made simpler by cost-benefit analysis.
2. The cost and benefit listing helps the analyst define each cost and benefit and review them later.
3. It determines whether the benefits outweigh the costs and whether they are financially sound and supportive.
4. Huge profit can be made through CBA, which makes things easier.
5. CBA is suitable for both large and small projects.
6. CBA decides by looking at the figures presented in the same unit.

6. Why we are using resiliency in a project?

Every project is designed for a certain period, and this design must ensure that it should serve its purpose without enduring any life and property. The cost-benefit analysis may be best used to describe the effectiveness of a project ensuring that there are alternatives in making investment decisions during the early phase of its life cycle. CBA helps in addressing the cost estimation, effectiveness, and efficiency in making an investment decision for a project. It also examines the measures and the methodology used to develop a CBA, addresses the accuracy and reliability of CBA, and identifies techniques available to support decision-making in the early phase of a program's lifecycle (Misuraca, 2014).

All power system infrastructure is vulnerable to severe weather; nevertheless, there are benefits to using solar PV as a robust power source, such as its distributed nature and lack of dependency on fuel. To fully take advantage of these benefits, solar arrays must be built and constructed in such a way that they have the best chance of surviving severe weather events and producing power thereafter. Because solar PV has grown and developed so quickly, and the industry has become so competitive, standard design, engineering, and construction practices are frequently neglected in the pursuit of lower upfront costs and faster project completion. It's also critical in maintaining a high standard of maintenance. New, stronger, and clearer standards in general might help to support and drive the PV industry toward more robust system design. Most significantly, while calculating the costs of surviving severe weather events, the costs of system damage or entire loss at the hands of a storm must be weighed against the high costs of system damage or total loss. This chapter helps in assisting a PV system in surviving a severe weather event or minimizing storm-related damages. We can't anticipate all of the failure modes and environmental circumstances that a system will experience during its lifespan; therefore adopting the recommendations in this study does not guarantee that the system will survive. The value resilient power systems can provide during and after severe weather occurrences is becoming increasingly significant. Severe weather-prone regions could benefit from resilient solar PV. As a resilient power source, solar PV offers several advantages. However, to be effective as a resilient power solution, it must be able to survive the weather event. It must be designed, installed, and maintained to a higher standard for its survival. Although doing so will almost certainly increase the cost, the advantages may outweigh the costs in many situations. In high-wind conditions, modules are subjected to periodic uplift, causing them to bend within their frames and away from their mounting fixtures. If the uplift pressure is high or persistent enough, the modules may be damaged or the glass top sheet may break.

6.1 Chances of occurrence of failure of a project

The probability of occurrence can let us know whether the work done would be enough or will they offer the protection/resilience for which they are designed. The only method that can offer such quantitative information on the prioritization of risk assessment and climate adaptation choices is by cost-benefit analysis.

The probability of failure of a project during its life time is given by Mays (2004).

If we are taking P as the probability of the occurrence of an event (It may be rainfall, earthquake, building structure, and bridges collapse), then,

$(1 - P)$ = Probability that the event will not happen $(1 - P)(1 - P)$
= Probability that the event will not occur in two succesive years $(1 - P)(1 - P)(1 - P)$
= Probability that the event will not occur in three succesive years
$(1 - P)^N$ = Probability that the event will not occur in a span of N succesive years

The risk (R) or the probability that the event will occur over a period of N years is given by, $R = 1 - (1 - P)^N$. The probability P is given by $P = \frac{1}{T_r}$. For return period T_r and various span of years N, the risk (R) that an event with a given return period will be equal to or exceeded over a span of N years.

When calculated, the risk that the event is reached or skipped over a while decreases with increasing return time. This effect is often used in the construction of large buildings. There is also an increase in costs by considering the structure of the building in the long run. However, this should be done to prevent disasters that cause loss of life and property (Chow et al., 1988).

Many codes of practice indicate that one of the reasons for choosing a return period of 50 years has been that the average lifetime of most buildings and structures is near 50 years. The use of better materials improves the life of buildings and structures. Generally, the drains or bridge culverts have a longer life (Proag & Proag, 2014).

The probability of occurrence can let us know whether the work done would be enough or will they offer the protection/resilience for which they are designed. The risk that the event is reached or skipped over a period of time decreases with increasing return time. The protection against such threats is concerned with system resilience. When the return period is a longer one, hazards are more serious (Elsworth & Van Geet, 2020).

In this study, we take an example of a solar PV plant. The study gives,

1. An early assessment of the increased costs associated with PV system events that occur under extreme weather conditions.

2. Encourage a greater consideration of the site environmental conditions and extreme weather events a PV system is likely to encounter over its operational lifetime.
3. Serve as a resource for developers, site operators, investors, and code and standard developers, among others, who are implementing systems under extreme weather conditions.
4. Installation of more resilient PV systems.
5. Lay the foundation for future work so that the costs of installing resilient PV systems may be more accurately estimated.

Solar photovoltaic (SPV) power offers several benefits as a resilient power source, including the capacity to provide power following a natural disaster. Although solar panels can withstand harsh weather, in some cases, systems are damaged and unable to generate power (Hotchkiss, 2016). PV systems must remain operational to act as resilient power sources. Building a system that is more likely to withstand a strong storm might be more expensive than building one that meets less stringent standards. Various practices can improve the chances of surviving a severe weather event of a PV system (Burgess & Goodman, 2018; Robinson, 2018). Strong foundations are required for solar PV plant designs to minimize high-wind damage. Another factor to consider is that solar power cannot generate electricity prior, during, or immediately after strong storms. How a solar PV project can be useful is provided below as an example.

7. Economic analysis of a photovoltaic system

The life cycle cost (LCC) of a 3.4-kW proposed grid connected SPV system was studied in severe weather conditions. The LCC of this PV system was carried out by present value method and embodied energy basis. All the capital investments, including the replacement cost of components and repair and maintenance cost, were converted to the present value by considering a discount rate of 10.62%. In addition to this discount rate, a price escalation rate of 5.72% was taken into consideration for repair and maintenance costs only due to the escalation of labor costs over the years. The project life was taken as 25 years and was considered by referring to the previous kinds of literature. The photovoltaic system is analyzed to find its viability for its cost economics. The LCC of a standalone solar PV system consists of total capital investment (*C*), its O&M, and replacement cost of inverter and components. The initial investment (*C*) is the sum of the cost of each part of the PV system, that is, PV array, inverter, control unit, battery and miscellaneous (electric cable, out house, etc.), including transportation and installation (Indicators—World Bank Data, n.d.). The operation and maintenance cost includes repair and scheduled maintenance costs per year. The following assumption and costs were considered for analysis. The discount rate, escalation rate, and energy tariff are taken as per the recommendation of the Odisha Electricity Regulatory Commission (OERC) for solar PV projects (Annual report-2012-13—OERC, 2014).

Three different types of investment costs are involved in PV power systems. These are:

Initial capital cost for installation of PV power system.

Recurring cost to be incurred every year for repair, operation, and maintenance.

Nonrecurring cost to be incurred at regular intervals for replacement of components after their life period.

7.1 Assumption taken for the PV system

Life period of PV project: 25 years.

Discount rate: 10.62% (Annual report-2012-13—OERC, 2014).

General inflation for O&M cost: 5.72% (Annual report-2012-13—OERC, 2014).

Salvage value: 0.

Maintenance cost: 2% of capital cost (Padmanathan et al., 2017) (calculated on capital cost excluding array structure and installation cost).

Life of battery: 5 years.

Life of inverter: 15 years (Fthenakis et al., 2011).

Electricity tariff: INR 5.30/kWh.

Solar energy tariff for grid connected PV system: INR 11.23/kWh for 12 years.

INR 6.81/kWh for 13 years (Annual report-2012-13—OERC, 2014).

7.2 Capital cost of the PV system

The total cost of installation of the PV system is given in Table 10.1.

7.3 Energy generation from PV system

Total life cycle energy generation by the PV system is $3.36\,\text{kWp} \times 4.62\,\text{kWh/m}^2\,\text{day} \times 365\,\text{days/year} \times 25\,\text{years} = 141{,}650\,\text{kWh}$ and energy generation per year is 5666 kWh.

Table 10.1 Total capital cost involved in a PV system.

Array cost (280 W × 12 nos): (3.36 kWp = 3.4 kWp)	INR 141,120
Battery cost (2 V, 340 Ah, 24 nos):	INR 136,000
Inverter cost:	INR 46,000
Controller cost:	INR 19,000
Array structure and installation:	INR 128,086
Total:	**INR 470,206**

7.4 Maintenance cost

Maintenance cost = 2% of INR (470,206 − 128,086) = INR 6842/year.

7.5 LCC of a grid-connected PV system

The annual revenue to be obtained from the sale of electricity to the state grid in case of a grid-connected PV system was computed by considering an energy tariff of INR 11.23/kWh for first 12 years and thereafter INR 6.81/kWh for the next 13 years.

The lifecycle benefit of PV system with domestic tariff and grid connected tariff is shown in Table 10.2.

7.6 Benefit-cost ratio (BCR)

If the benefit-cost ratio is greater than 1, the project is considered a profitable venture. The benefit-cost ratio at 10.62% discount rate is:

Table 10.2 Lifecycle benefit of the PV system (project life: 25 years).

	Grid-connected system	
Parameters	**1st 12 years**	**Next 13 years**
Annual energy generation, kWh/year	5666 (supply to grid)	5666 (supply to grid)
Energy tariff, INR/kWh	11.23	6.81
Annual revenue, INR/year	63,629	38,585
Multiplication factor	$[(1+r)^n - 1]/r(1+r)^n = 6.61$ for $n = 12$ years	$[(1+r)^n - 1]/r(1+r)^n$ 6.88 for $n = 13$ years
Present value of benefit, INR	420,689 (at base year 2014) (PV_{12}) 79078 (at base year 2014) (PV_{13}) by multiplication factor $1/(1+r)^n$ ($PV_{12} + PV_{13}$) 499,767	265,497 at 12th year

As the BCR is less than 1, the project is considered to be a nonprofitable project.

Grid connected system BCR = present value of benefit/present value of cost = 4, 99,767/4, 48,569 = 1.11.

The present values of all revenue generated (benefit) and the present values of all investment (cost) of a similar grid-connected solar PV system at a discount rate of 10.62% will be INR 448,569.

As the BCR is more than 1, the project is considered to be a profitable project based on an economic point of view. So, grid-connected PV system is a profitable one.

8. Conclusion

Cost-benefit analysis is based on future predictions that may or may not come true. As a result, CBA models include a significant amount of uncertainty. CBA models can account for uncertainty and variability like natural disasters and fluctuation in price. Sensitivity analysis is used in a CBA model to evaluate the model's underlying assumptions and see how changing these assumptions affects the outcome of the project. This chapter also provides recommendations for assisting a PV system in surviving a severe weather event or minimizing damages caused due to storm. Solar PV that is resistant to severe weather might help regions that are prone to severe weather. However, to be useful as a resilient power solution, the system must be able to withstand a weather event.

References

Annual report-2012-13—OERC. (2014). http://www.orierc.org/ORDERS%5C2012%5CC-01-2012.PDF.

Burgess, C., & Goodman, J. (2018). *Solar under storm: Select best practices for resilient ground-mount PV systems with hurricane exposure*. Boulder, CO: Rocky Mountain Institute.

Chow, V. T., Maidment, D. R., & Mays, L. W. (1988). *Applied hydrology*. McGraw Hill.

Cost-benefit analysis: Advantages, limitations, examples, and relevance. (2019). https://www.marketing91.com/cost-benefit-analysis/.

Elsworth, J., & Van Geet, O. (2020). *Solar photovoltaics in severe weather: Cost considerations for storm hardening PV systems for resilience (No. NREL/TP-7A40-75804)*. Golden, CO: National Renewable Energy Laboratory.

Fthenakis, V., Frischknecht, R., Raugei, M., Kim, H. C., Alsema, E., Held, M., & Wild-Scholten, & de. (2011). *Methodology guidelines on life cycle assessment of photovoltaic electricity*. IEA PVPS Task.

Goel, S., Sharma, R., & Jena, B. (2021). Life cycle cost and energy assessment of a 3.4 kWp rooftop solar photovoltaic system in India. *International Journal of Ambient Energy*, 1–11. https://doi.org/10.1080/01430750.2021.1913221.

Hotchkiss, E. (2016). *Bonus module: Using solar for resilience. City and county solar PV training program.*

Indicators—World Bank Data. (n.d.). Retrieved from: http://dataworldbank.org/indicator. (Accessed 28 October 2017).

Landau, P. (2021). *Cost benefit analysis for projects A step-by-step guide.* https://www.projectmanager.com/blog/cost-benefit-analysis-for-projects-a-step-by-step-guide.

Lin, C. H., Hsieh, W. L., Chen, C. S., Hsu, C. T., Ku, T. T., & Tsai, C. T. (2011). Financial analysis of a large-scale photovoltaic system and its impact on distribution feeders. *IEEE Transactions on Industry Applications, 47*(4), 1884 1891. https://doi.org/10.1109/TIA.2011.2154292.

Makowsky, M. D., & Wagner, R. E. (2009). From scholarly idea to budgetary institution: The emergence of cost-benefit analysis. *Constitutional Political Economy, 20*(1), 57–70. https://doi.org/10.1007/s10602-008-9051-7.

Mays, L. W. (2004). *Water resources engineering.* USA: John Wiley.

Misuraca, P. (2014). *The effectiveness of a costs and benefits analysis in making Federal Government decisions: A literature review.*

Padmanathan, K., Govindarajan, U., Ramachandaramurthy, V. K., & Sudar Oli Selvi, T. (2017). Multiple criteria decision making (MCDM) based economic analysis of solar PV system with respect to performance investigation for Indian market. *Sustainability, 9* (5), 820. https://doi.org/10.3390/su9050820.

Proag, S.-L., & Proag, V. (2014). The cost benefit analysis of providing resilience. *Procedia Economics and Finance, 361–368.* https://doi.org/10.1016/s2212-5671(14)00951-4.

Robinson, G. (2018). *Solar photovoltaic systems in hurricanes and other severe weather.* https://www.energy.gov/sites/prod/files/2018/08/f55/pv_severe_weather.pdf.

Sowe, S., Ketjoy, N., Thanarak, P., & Suriwong, T. (2014). Technical and economic viability assessment of PV power plants for rural electrification in the Gambia. *Energy Procedia, 52*, 389–398. https://doi.org/10.1016/j.egypro.2014.07.091.

Index

Note: Page numbers followed by *f* indicate figures and *t* indicate tables.

G

H

I

L

M

N

O

P

Q

R

S

T

Printed in the United States
by Baker & Taylor Publisher Services